Fault Detection and Fault-Tolerant Control for Nonlinear Systems

Linlin Li

Fault Detection and Fault-Tolerant Control for Nonlinear Systems

 Springer Vieweg

Linlin Li
Duisburg, Germany

Dissertation University of Duisburg-Essen, Germany, 2015

ISBN 978-3-658-13019-0   ISBN 978-3-658-13020-6 (eBook)
DOI 10.1007/978-3-658-13020-6

Library of Congress Control Number: 2016931840

Springer Vieweg
© Springer Fachmedien Wiesbaden 2016

This Springer Vieweg imprint is published by Springer Nature
The registered company is Springer Fachmedien Wiesbaden GmbH

To my parents and Wanjun

Preface

Associated with increasing demands on system safety and reliability, fault detection (FD) and fault-tolerant control (FTC) have attracted considerable attention in both research and application fields. Due to the continuously increasing system automation, integration and complexity degrees, industrial processes are typically nonlinear. Therefore, developing FD and FTC approaches for nonlinear systems belong definitely to the most remarkable and challenging topics.

This work is devoted to address the analysis and design issues of observer-based FD and FTC for nonlinear systems. In the first part of the thesis, the configuration of nonlinear observer-based FD systems is formulated by parameterizing the residual generators. Based on the parameterization form, the nonlinear observer-based FD systems are parameterized as well as the threshold settings. Furthermore, the existence conditions of the nonlinear observer-based FD systems are studied to gain a deeper insight into the construction of the FD systems.

The second part of the work focuses on the developments of FD schemes by dealing with the proposed conditions with the aid of the Takagi-Sugeno (T-S) fuzzy dynamic modelling techniques. To further improve the FD performance, an alternative fuzzy observer-based approach is proposed by making use of the knowledge provided by the fuzzy models of each local region and weighting the local residual signal by means of different weighting factors. This is motivated by the fact that unlike linear systems with unified dynamics over the whole working range, the local behavior of nonlinear systems can be significantly different.

With the FD system at hand, it is important to re-configure the controller to maintain or recover the system operations after an alarm is given. For this purpose, the third part of the work is dedicated to two FTC configurations for a class of nonlinear systems. The proposed architectures provide an integrated solution that has advantages to make the plant maintenance, repair and operations easier to handle. Finally, the derived FD and FTC approaches are verified by two benchmark processes. The application results demonstrate the effectiveness of the developed methods.

This work was done while the author was with the institute for Automatic Control and Complex Systems (AKS) at the University of Duisburg-Essen. I would like to give the most sincere thanks to Prof. Dr.-Ing. Steven X. Ding for his guidance to my scientific research work. I am very grateful for all his help, encouragements and insight discussions on this work during the past three years. My sincere appreciation must also go to Prof. Ying Yang for her valuable guidance and discussions on this thesis. I am very grateful for her consistent support and encouragements. I would also like to thank Prof. Jianbin Qiu for all his valuable guidance, discussions and cooperation on the fuzzy fault detection works. I am very grateful for his consistent patience and support.

I would like to thank Kai, Zhiwen, Tim K., Yuri, Hao, Minjia and Svenja for their valuable review and suggestions to my dissertation. My thanks also go to all the AKS colleagues, Shane, Changchen, Sihan, Haiyang, Tim D., Chris, Ali, Christoph, Sabine, Dr. Köppen-Seliger, Eberhard, Klaus, Göbel, Dr. Zhao, Dr. Yang, Prof. Peng and Prof. Lei for their valuable discussion and helpful suggestions.

My special thanks go to my group mate Yong for all his valuable discussions and cooperation. I would like to thank Dongmei for her great support and company. Moreover, I am also grateful to my former colleagues and friends, Hongli, Bo and Shouchao for their valuable support and encouragements. I would also like to thank my friends, Yao and Jing, for their valuable support.

Finally, I would like to dedicate this work to my family, especially my parents, my sister and my brother for understanding and supporting me in whatever I decide to do - especially my husband, Wanjun, for his patience and support.

Linlin Li

Contents

List of Figures

List of Tables

List of Notations

Abbreviations

Abbreviation	Expansion
CSTH	continuous stirred tank heater
EIMC	extended internal model control
FD	fault detection
FDD	fault detection and diagnosis
FDF	fault detection filter
FDI	fault detection and isolation
FE	fault estimation
FTC	fault-tolerant control
GIMC	generalized internal model control
HJI	Hamilton-Jacobi inequality
H-PRIO	high-priority
IMC	internal model control
IOS	input-to-output stable
LCF	left coprime factorization
LMI	linear matrix inequality
L-PRIO	low-priority
LPV	linear parameter-varying
LTI	linear time-invariant
NCS	networked control system
PID	proportional-integral-derivative
PD	proportional-derivative
PI	proportional-integral
RCF	right coprime factorization
ROS	robust-to-output stable
RK	Runge-Kutta
SKR	stable kernel representation
SIR	stable image representation
T-S	Takagi-Sugeno

Mathematical notations

Notation	Description
$\mathcal{U}$	a vector space of functions from a time domain to a Euclidean vector space
$\mathcal{U}^s$	the stable subset of $\mathcal{U}$
$\Sigma^{\mathbf{x}_0} : \mathcal{U} \to \mathcal{Y}$	An operator Σ with input signal space $\mathcal{U}$, output signal space $\mathcal{Y}$ and initial condition $\mathbf{x}_0$
$\Sigma_a^{\mathbf{x}_a,0} \circ \Sigma_b^{\mathbf{x}_b,0}$	the cascade connection of two systems $\Sigma_a^{\mathbf{x}_a,0} : \mathcal{U} \to \mathcal{Y}$ and $\Sigma_b^{\mathbf{x}_b,0} : \mathcal{Y} \to \mathcal{Z}$
$\mathcal{RH}_\infty$	the set of all stable transfer matrices
$\forall$	for all
$\in$	belong to
$\Longrightarrow$	imply
$\Longleftrightarrow$	equivalent to
$\|\cdot\|$	Euclidean norm of a vector
$\|\cdot\|_2$	$\mathcal{L}_2$ norm of a signal
$\hat{\mathbf{x}}$	estimate of the state vector $\mathbf{x}$
$\mathbf{x}$	a vector
$\mathbf{X}$	a matrix
$\mathbf{X}^T$	transport of $\mathbf{X}$
$\mathbf{X}^{-1}$	inverse of $\mathbf{X}$
$\mathbf{X} > 0$	$\mathbf{X}$ is positive definite matrix
$\mathcal{R}^n$	space of n-dimensional vectors
$\mathcal{R}^{n \times m}$	space of n by m matrices
$\mathrm{Sym}\{\mathbf{X}\}$	$\mathbf{X} + \mathbf{X}^T$
$\mathcal{B}_\delta$	space of the vector $\mathbf{x} \in \mathcal{R}^n$ satisfying $\|\mathbf{x}\| \leq \delta$ for some $\delta > 0$
$V_\mathbf{x}(\mathbf{x})$	$\frac{\partial V(\mathbf{x})}{\mathbf{x}}$
$J(\mathbf{r})$	evaluation function
J_{th}	threshold
$\mathbf{u}_\tau$	the truncation of $\mathbf{u}$ at τ, i.e. $\mathbf{u}_\tau(t) = \mathbf{u}(t)$ if $t \leq \tau$, and $\mathbf{u}(t) = \mathbf{0}$ if $t > \tau$
$\mathbf{I}_m$	m by m identity matrix
$\beta(\cdot) \in \mathcal{K}$	$\beta : \mathcal{R}_+ \to \mathcal{R}_+$ is continuous, strictly increasing, and satisfies $\beta(0) = 0$
$\beta(\cdot) \in \mathcal{K}_\infty$	$\beta \in \mathcal{K}$, and in addition, $\lim_{s \to \infty} \beta(s) = \infty$
$\beta(\cdot) \in \mathcal{L}$	β is continuous, strictly decreasing, and satisfies

| $\phi(s,t) \in \mathcal{KL}$ | $\lim_{s \to \infty} \beta(s) = 0$
 for each fixed t the function is of class $\mathcal{K}$ and for each fixed s it is of class $\mathcal{L}$ |

1 Introduction

1.1 Motivation of the Work

This thesis deals with observer-based fault detection (FD) and fault-tolerant control (FTC) issues for nonlinear systems. The motivation of working on this topic will be explained consecutively.

Why FD and FTC?

Thanks to the rapid development of computer technique, electronics and information technology, modern industrial processes are generally becoming more and more complex. For such systems, the safety and reliability issues are of significant importance, since fault or failure may result in disastrous consequences and hazards for personnel, plant and environment, in particular when the systems are embedded in safety relevant plants like aeroplanes, vehicles, networks, robots or nuclear power plants. Few such incidents in the history are listed below:

- The worst nuclear power plant accident occurred in Chernobyl in 1986 resulting in 31 casualties of the reactor staff and emergency workers directly. In addition, around 600,000 people are exposed to the highest levels of radiation and the total economic cost of the disaster amounts to approximately 18 billions dollars [146].

- The American Airlines Flight 191 crashed after takeoff from Chicago, which killed 258 passengers, 13 crew members and two people on the ground. The accident was caused by the separation of the engine from the aircraft and the poor fault maintenance procedure.

In addition, it has been reported in a survey [135] that 20 billion dollars are lost per year due to poor abnormal event management in petrochemical industries. Thus, it is of great application interests to monitor the operation of the plants and react to the potential problems in a timely manner, so as to enhance the safety and reliability of the processes. Strongly driven by these observations, numerous methods have been proposed in the past decades. Among involved studies, FD and FTC have received considerably increasing attention both in the research and application domains.

Why Observer-based FD?

Driven by this industrial safety and reliability demand, great efforts have been made to strengthen the research on FD approaches. Roughly speaking, a fault is caused by malfunctions or deviations which prevent the fulfillment of the specified functionality and reside temporarily or permanently in system components, actuators and sensors. The traditional FD techniques can be classified into three categories: hardware redundancy schemes, signal processing based approaches and model-based (analytical redundancy) schemes.

The core of hardware redundancy schemes lies in reconstructing a redundant process by using the identical hardware components [30]. By continuously comparing the output of the process and the one of its redundant process, fault detection and direct fault isolation can be achieved. This scheme promises a high reliability and thus can be applied in safety relevant plants. However, the high economical costs for the redundant hardware components limit its application.

The basic idea of signal processing based approaches consists in collecting and analyzing the measurable data of the process to realize the detection of the fault. In fact, today's automatic control systems are generally equipped with an excellent information infrastructure that allows an automatized process data collection, archiving and efficient off- and on-line data processing, which drives this approach. However, in general, the signal processing techniques are mainly valid for linear stationary processes [53].

Model-based FD approaches, initiated in the 1970s [12, 67], have established themselves as an attractive research area and served as the main methodologies in the area of automatic control systems during the past four decades [25, 30, 117, 14, 94]. It is also called analytical redundancies based approach, since a software redundant model is running in parallel to the process, as shown in Fig. 1.1. A standard model-based FD system is composed of a residual generator, an evaluation

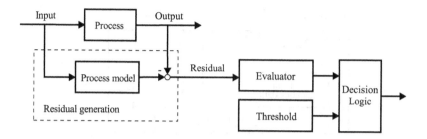

Figure 1.1: Schematic description of model-based fault detection configuration

function and a detection logic with a threshold [30]. The difference between the output of the process and the redundant model is called residual signal. Roughly speaking, residual generation is equivalent with building analytical redundancy. Residual evaluation serves the purpose of making a right decision for fault detection. The application of the analytical model-based design scheme may result in a significant reduction in engineering costs compared with the hardware redundancy approach. Model-based FD approaches can be further classified into observer-based FD approaches, parameter identification based approaches and parity space approaches. In the past decades, there has been increased emphasis on the development of observer techniques, associated with the high demands on the reliable estimation of unmeasurable variables in industrial and commercial processes [92, 103]. Driven by this observation, observer-based FD approaches have been extensively investigated and built the main stream of model-based FD approaches in recent years [160, 168, 33].

Why Active FTC?

It is well-known that the FTC techniques are indispensable tools to deal with the possible performance degradation. During the past two decades, numerous approaches have been proposed to the model-based fault-tolerant control (FTC) [19, 154, 152, 13, 85, 10, 90, 150, 115], aiming at improving reliability of automatic control systems. Roughly speaking, FTC techniques can be classified into two catogaries: passive FTC and active FTC. The basic idea behind passive FTC schemes lies in the fact that the information of possible faults is taken into account in the

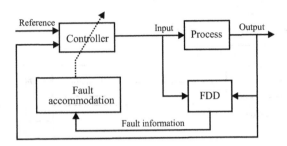

Figure 1.2: Active fault-tolerant control architecture

control law design a priori [13, 85, 90, 150, 115]. In this context, most of research efforts have been made on robust/reliable controller design [85, 90, 150, 115]. The active FTC approaches, on the other hand, rely on the information of fault delivered by the fault detection and diagnosis (FDD) module. A schematic description of this active FTC architecture is given in Fig. 1.2. Assume that a fault would cause a change in the system configuration. The FDD module delivers an estimate of the fault [63, 158], which will then activate an online reconfiguration and computation of control law [64]. Active FTC is one of the focuses of this thesis.

Why FD and FTC for Nonlinear Systems?

Linear control theory is a well established model-based framework, in which rich methods and tools are available for designing automatic control systems, including observer-based FD and FTC approaches. Reviewing the publications on FD and FTC studies reveals that the framework of designing observer-based FD and FTC for linear systems has been well established [30, 14, 93, 108]. However, associated with the continuously increasing system automation, integration and complexity degrees, industrial processes are typically nonlinear. Observer-based FD and FTC issues for nonlinear systems have received considerable attention in recent years [52, 88, 143]. Among involved studies, focuses are on dealing with different types of special nonlinear systems, for instance, affine nonlinear systems [1, 109, 69, 143], Lipschitz nonlinear systems [27, 163, 87], nonlinear networked-control systems (NCSs) [95, 96] and linear systems with nonlinear uncertainties [147, 162], etc.

A thorough literature review on the nonlinear FD works has revealed that, up to now, little attention has been paid to the integrated design of nonlinear observer-based FD systems, which consists of an observer-based residual generator, an evaluation function and a decision logic with an embedded threshold [1, 72, 151]. Most studies mainly concentrated on the design schemes of observer-based residual generators. Moreover, the situation remains the same as dealing with the integrated design of observer-based FD systems for general nonlinear processes. In particular, most of research efforts on nonlinear FD have been made on the FD system design but only few on the analysis of fundamental properties, for instance, the design and existence conditions of nonlinear observer-based FD systems.

On another research frontier, rapidly growing interest has been devoted to Takagi-Sugeno (T-S) fuzzy-model-based analysis and synthesis techniques [40, 116, 126, 18, 131]. In particular, it has been demonstrated that a T-S fuzzy dynamic model, being composed by a number of local models, can well describe the global behaviors of a highly complex nonlinear process or even nonanalytic system [126, 18, 131]. As a result, a framework of designing T-S fuzzy-model-based controllers and observers for nonlinear systems has been well established and many methods in this framework have been successfully applied in practice [131, 74, 89, 60, 48, 155, 101, 128, 26, 142]. Encouraged by these successful results, intensive attention has been conducted attempting to apply the T-S fuzzy-model-based technique to deal with FD problems for affine nonlinear systems [49, 37, 166, 144, 20, 167]. Nevertheless, there have been few results on the integrated design of fuzzy-observer-based FD systems for complex nonlinear processes.

In control theory, system parameterization is an essential form for system analysis and optimization. Inspired by the well-known Youla parameterization of all stabilizing controllers [169] using factorization technique, a parameterization form of residual generators for LTI systems has been first proposed in [35], which plays an important role in linking the residual generation and evaluation as well as their optimization in the observer-based FD framework [30]. An extension of this parameterization form to nonlinear observer-based residual generators has been addressed in the study on optimizing residual evaluation and threshold setting [1]. Despite this work, there have been very few efforts dedicated to the systematic study on the parameterization of nonlinear FD systems.

Moreover, controller parameterization plays an essential role in optimizing system performance for linear systems. Towards the design for performance and robustness, [171] proposed the so-called generalized internal model control (GIMC) structure, which enables system performance and robustness to be designed separately. Based on it, [34] demonstrates that all stabilizing controllers can be equivalently realized in the observer form. The residual signal, also known as the estimated output error in [171] , not only benefits the study of fault detection and diagnosis, but also enhances the design of feedback controllers for stability and robustness issues. Based on this strategy, the extended internal model control (EIMC) structure has been proposed with an observer-based residual generator parallel running in the control loop. Compared with linear FTC, FTC for nonlinear systems is still an open and challenging topic. So far, there are few efforts dedicated to the observer-based realizations of controller parameterization and FTC architectures for nonlinear systems. In particular, the separate design for system stability performance and fault tolerance seems to be left out of investigation. The aforementioned reasons motivate the study on FD and FTC schemes for nonlinear systems.

1.2 Objectives of the Work

The main objective of this thesis is to analyze and design observer-based FD and FTC approaches for nonlinear systems. To be specific, the tasks of this thesis are stated as follows:

- formulate the configuration of nonlinear observer-based FD systems by parameterizing the residual generators. Based on the parameterization form, the nonlinear observer-based FD systems are parameterized as well as the threshold settings.

- study an essential analysis issue, the existence conditions of observer-based FD systems for nonlinear processes, to gain a deeper insight into the fundamental properties of nonlinear observer-based FD systems.

- propose the integrated design schemes of nonlinear observer-based FD systems by dealing with the proposed conditions with the aid of T-S fuzzy dynamic modelling techniques.

- develop a real-time nonlinear observer-based FD approach to realize an early detection of potential faults.

- propose a weighted fuzzy observer-based FD approach for nonlinear systems aiming at improving the FD performance of the conventional approaches.

- formulate two configurations for active nonlinear FTC systems to realize the separate design of system stability and fault tolerance.

1.3 Outline of the Thesis

This thesis consists of nine chapters. The major objectives and contributions of each chapter are briefly summarized as follows.

Chapter 1: Introduction

This chapter presents the motivations, objectives, contributions and the organization of this thesis.

Chapter 2: Overview of FD and FTC Technologies

This chapter is devoted to the overview of FD and FTC methodologies for both linear and nonlinear systems. Based on coprime factorization techniques, the configurations of observer-based FD systems and FTC for linear systems are reviewed. Then the technical descriptions for different types of nonlinear systems are provided.It is followed by a briefly summation of the basic techniques for nonlinear FD and FTC techniques.

Chapter 3: Configuration of Nonlinear Observer-based FD Systems

This chapter addresses the analysis and integrated design of observer-based FD systems for nonlinear processes. To gain a deeper insight into the observer-based FD framework, definitions and existence conditions for observer-based FD systems for nonlinear processes are investigated first. Then the parametrization issues of nonlinear observer-based fault detection (FD) systems are studied. This study consists of two steps. In the first step, with the aid of nonlinear factorization and input-output operator techniques, it is proved that any stable residual generator can be parameterized by a cascade connection of the process kernel representation and a post-filter that represents the parameter system. In the second step, based on the state space representation of the parameterized residual generators, different classes of observer-based FD systems are investigated. This leads to the parametrization of the threshold settings for both classes of FD systems and, associated with them, to the characterization of the existence conditions.

Chapter 4: Design of $\mathcal{L}_2$ Nonlinear Observer-based FD Systems

In this chapter, an integrated design scheme of observer-based FD system is proposed for general nonlinear industrial processes with the aid of T-S fuzzy dynamic modelling technique. To be specific, the nonlinear systems are first approximated by a class of generalized T-S fuzzy models. Then, the universal T-S fuzzy observer-based residual generator is studied along the lines of $\mathcal{L}_2$ stability theory. Moreover, an integrated observer-based FD scheme is proposed with an embedded dynamic threshold, which is generated to meet the online and real-time fault detection requirements from industrial processes. In the end, the effectiveness of the proposed method is verified by a numerical example.

Chapter 5: Design of $\mathcal{L}_\infty/\mathcal{L}_2$ Nonlinear Observer-based FD Systems

In this chapter, a real-time observer-based FD approach for a general type of nonlinear systems in the presence of external disturbances is studied. To this end, in the first part of this chapter, the definition and the design conditions for an $\mathcal{L}_\infty/\mathcal{L}_2$ type of nonlinear observer-based FD systems are studied. In the second part, the integrated design of the $\mathcal{L}_\infty/\mathcal{L}_2$ observer-based FD systems is addressed by applying T-S fuzzy dynamic modelling technique as the solution tool. In particular, this fuzzy observer-based FD approach is developed via piecewise Lyapunov functions and, can be applied to the case that the premise variables of the FD system are non-synchronous with the premise variables of the fuzzy model for the plant.

Chapter 6: Design of Weighted Fuzzy Observer-based FD Systems for Discrete-Time Nonlinear Processes

This chapter is devoted to the analysis and integrated design of $\mathcal{L}_2$ observer-based FD systems for discrete-time nonlinear industrial processes.

To gain a deeper insight into this FD framework, the existence condition is introduced first. Then, an integrated design of $\mathcal{L}_2$ observer-based FD approach is carried out by solving the proposed existence condition with the aid of T-S fuzzy dynamic modelling technique and piecewise-fuzzy Lyapunov functions. Most importantly, a weighted piecewise-fuzzy observer-based residual generator is proposed aiming at achieving an optimal integration of residual evaluation and threshold computation into FD systems. The core of this approach is to make use of the knowledge provided by fuzzy models of each local region and then weight the local residual signal individually, instead of constantly, by means of weighting factors. In comparison with the standard norm-based fuzzy observer-based FD methods, the proposed scheme may lead to a significant improvement of the FD performance. The effectiveness of the proposed method is demonstrated using a numerical example.

Chapter 7: FTC Configurations for Nonlinear Systems

This chapter investigates the configurations of fault-tolerant controllers for affine nonlinear systems. The design philosophy is highlighted by the observer and residual generator based controller parametrization and the integration of a fault diagnosis system. To be specific, a novel interpretation for the design of fault-tolerant controllers is first introduced with the combination of any stabilizing controller and a residual-driven dynamic compensator. It allows us to attain the separate design of the performance and fault tolerance. Under such circumstances, the whole design procedures finally realize the maintenance and life-circle management of affine nonlinear systems. In the end, a design scheme of the FTC framework is proposed.

Chapter 8: Application to Benchmark Processes

In this chapter, the methods proposed in Chapters 3-7 are tested on two industrial benchmark processes, the laboratory setup of continuous stirred tank heater (CSTH) process and three-tank system.

Chapter 9: Conclusions and Future Work

This chapter concludes the thesis and discusses the future scope.

2 Overview of FD and FTC Technology

This chapter reviews the FD and FTC technology for both LTI and nonlinear systems, respectively. First, the configurations of observer-based FD and FTC systems for LTI processes are examined, which motivate the investigations on nonlinear FD and FTC configurations in this thesis. Then, a brief description for different types of nonlinear systems is presented. It is followed by the state-of-the-art and the basic methodologies of fault detection and isolation (FDI) and FTC techniques for nonlinear systems.

2.1 FD and FTC Configuration for LTI Systems

Let the minimal state space representation of LTI systems be

$$\mathbf{G} : \begin{cases} \dot{\mathbf{x}} = \mathbf{A}\mathbf{x} + \mathbf{B}\mathbf{u} \\ \mathbf{y} = \mathbf{C}\mathbf{x} + \mathbf{D}\mathbf{u} \end{cases} \tag{2.1}$$

where $\mathbf{x} \in \mathcal{R}^{k_x}$, $\mathbf{u} \in \mathcal{R}^{k_u}$ and $\mathbf{y} \in \mathcal{R}^{k_y}$ represent the state, measured input and output vectors; $\mathbf{A}$, $\mathbf{B}$, $\mathbf{C}$ and $\mathbf{D}$ are system matrices of appropriate dimensions. The corresponding transfer function for system (2.1) is

$$\mathbf{G}(s) = \mathbf{C}(s\mathbf{I} - \mathbf{A})^{-1}\mathbf{B} + \mathbf{D}. \tag{2.2}$$

In linear FD framework [30], it is generally assumed that the influence of disturbances and faults is modelled by extending (2.1) to

$$\mathbf{G}_{\mathbf{w,d}} : \begin{cases} \dot{\mathbf{x}}(t) = \mathbf{A}\mathbf{x}(t) + \mathbf{B}\mathbf{u}(t) + \mathbf{E}_{\mathbf{w}}\mathbf{w}(t) + \mathbf{E}_{\mathbf{d}}\mathbf{d}(t) \\ \mathbf{y}(t) = \mathbf{C}\mathbf{x}(t) + \mathbf{D}\mathbf{u}(t) + \mathbf{F}_{\mathbf{w}}\mathbf{w}(t) + \mathbf{F}_{\mathbf{d}}\mathbf{d}(t) \end{cases} \tag{2.3}$$

where $\mathbf{w} \in \mathcal{R}^{k_w}$ and $\mathbf{d} \in \mathcal{R}^{k_d}$ represents the fault to be detected and unknown disturbance vector, respectively; and $\mathbf{E_w}, \mathbf{E_d}, \mathbf{F_w}$ and $\mathbf{F_d}$ are known matrices of appropriate dimensions.

Next, the coprime factorization techniques for linear systems are considered, which play an essential role in formulating the FD and FTC configurations for LTI systems.

2.1.1 Coprime Factorization Techniques

The following definitions are necessary to introduce linear coprime factorization techniques.

Definition 2.1. *(Right-coprime factorization (RCF) [169]) A factorization* $\mathbf{G}(s) = \mathbf{N}(s)\mathbf{M}^{-1}(s)$ *is said to be an RCF of* $\mathbf{G}(s)$ *if (i)* $\mathbf{N}(s) \in \mathcal{RH}_\infty$ *and* $\mathbf{M}(s) \in \mathcal{RH}_\infty$ *and (ii) there exists* $\mathbf{Y}(s) \in \mathcal{RH}_\infty$ *and* $\mathbf{X}(s) \in \mathcal{RH}_\infty$ *such that*

$$\mathbf{Y}(s)\mathbf{M}(s) + \mathbf{X}(s)\mathbf{N}(s) = \mathbf{I}. \tag{2.4}$$

Definition 2.2. *(Left-coprime factorization (LCF) [169]) A factorization* $\mathbf{G}(s) = \hat{\mathbf{M}}^{-1}(s)\hat{\mathbf{N}}(s)$ *is said to be an LCF of* $\mathbf{G}(s)$ *if (i)* $\hat{\mathbf{N}}(s) \in \mathcal{RH}_\infty$ *and* $\hat{\mathbf{M}}(s) \in \mathcal{RH}_\infty$ *and (ii) there exists* $\hat{\mathbf{Y}}(s) \in \mathcal{RH}_\infty$ *and* $\hat{\mathbf{X}}(s) \in \mathcal{RH}_\infty$ *such that*

$$\hat{\mathbf{N}}(s)\hat{\mathbf{X}}(s) + \hat{\mathbf{M}}(s)\hat{\mathbf{Y}}(s) = \mathbf{I}. \tag{2.5}$$

Given RCF and LCF of $\mathbf{G}(s)$, as well as $\mathbf{X}(s), \mathbf{Y}(s), \hat{\mathbf{X}}(s), \hat{\mathbf{Y}}(s)$, it is known from [169] that the generalized double Bezout identity holds

$$\begin{aligned}
&\begin{bmatrix} \mathbf{Y}(s) & \mathbf{X}(s) \\ -\hat{\mathbf{N}}(s) & \hat{\mathbf{M}}(s) \end{bmatrix} \begin{bmatrix} \mathbf{M}(s) & -\hat{\mathbf{X}}(s) \\ \mathbf{N}(s) & \hat{\mathbf{Y}}(s) \end{bmatrix} \\
&= \begin{bmatrix} \mathbf{M}(s) & \hat{\mathbf{X}}(s) \\ -\mathbf{N}(s) & \hat{\mathbf{Y}}(s) \end{bmatrix} \begin{bmatrix} \mathbf{Y}(s) & -\mathbf{X}(s) \\ \hat{\mathbf{N}}(s) & \mathbf{M}(s) \end{bmatrix} = \begin{bmatrix} \mathbf{I} & \mathbf{0} \\ \mathbf{0} & \mathbf{I} \end{bmatrix}.
\end{aligned} \tag{2.6}$$

Assuming that system (2.1) is controllable and observable, the transfer functions in (2.6) can be constructed as follows

$$\mathbf{M}(s) = \mathbf{F}(s\mathbf{I} - \mathbf{A} - \mathbf{BF})^{-1}\mathbf{B} + \mathbf{I}$$
$$\mathbf{N}(s) = (\mathbf{C} + \mathbf{DF})(s\mathbf{I} - \mathbf{A} - \mathbf{BF})^{-1}\mathbf{B} + \mathbf{D}$$

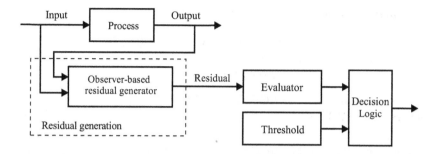

Figure 2.1: Observer-based fault detection configuration

$$\mathbf{Y}(s) = \mathbf{F}(s\mathbf{I} - \mathbf{A} - \mathbf{BF})^{-1}\mathbf{L}$$

$$\mathbf{X}(s) = (\mathbf{C} + \mathbf{DF})(s\mathbf{I} - \mathbf{A} - \mathbf{BF})^{-1}\mathbf{L} + \mathbf{I}$$

$$\hat{\mathbf{X}}(s) = -\mathbf{F}(s\mathbf{I} - \mathbf{A} + \mathbf{LC})^{-1}(\mathbf{B} - \mathbf{LD}) + \mathbf{I}$$

$$\hat{\mathbf{Y}}(s) = \mathbf{F}(s\mathbf{I} - \mathbf{A} + \mathbf{LC})^{-1}\mathbf{L}$$

$$\hat{\mathbf{N}}(s) = \mathbf{C}(s\mathbf{I} - \mathbf{A} + \mathbf{LC})^{-1}(\mathbf{B} - \mathbf{LD}) + \mathbf{D}$$

$$\hat{\mathbf{M}}(s) = -\mathbf{C}(s\mathbf{I} - \mathbf{A} + \mathbf{LC})^{-1}\mathbf{L} + \mathbf{I} \tag{2.7}$$

where $\mathbf{F}$, $\mathbf{L}$ are matrices of appropriate dimensions that ensure the stability of $\mathbf{A} + \mathbf{BF}$ and $\mathbf{A} - \mathbf{LC}$. Details can be found in [169].

2.1.2 The Configuration of Observer-based FD Systems

Fig. 2.1 shows the architecture of a standard observer-based FD system, which consists of an observer-based residual generator, a residual evaluator and a decision block with an embedded threshold [30].

Observer-based Residual Generator

To achieve a successful fault detection, residual generators play an key role in generating the residual signal that is sensitive to fault variables.

Definition 2.3. *(Observer-based residual generator [30]) An observer-based residual generator for (2.3) is a system*

$$\mathbf{r}(s) = \mathbf{K}(s) \begin{bmatrix} \mathbf{u}(s) \\ \mathbf{y}(s) \end{bmatrix} \tag{2.8}$$

such that (i) for any input $\mathbf{u}$,

$$\lim_{t \to \infty} \mathbf{r}(t) = \mathbf{0}, \quad \text{if } \mathbf{w} = \mathbf{0}, \mathbf{d} = \mathbf{0} \tag{2.9}$$

(ii) for some fault in the system, $\mathbf{r}(t) \neq \mathbf{0}$.

It is worth noting that LCF $\mathbf{G}(s) = \hat{\mathbf{M}}^{-1}(s)\hat{\mathbf{N}}(s)$ implies a residual generator for (2.3) as

$$\mathbf{r}(s) = \hat{\mathbf{M}}(s)\mathbf{y}(s) - \hat{\mathbf{N}}(s)\mathbf{u}(s) = \begin{bmatrix} -\hat{\mathbf{N}}(s) & \hat{\mathbf{M}}(s) \end{bmatrix} \begin{bmatrix} \mathbf{u}(s) \\ \mathbf{y}(s) \end{bmatrix}. \tag{2.10}$$

Together with (2.7), the state space representation of (2.10) is given by

$$\begin{aligned}
\dot{\hat{\mathbf{x}}}(t) &= \mathbf{A}\hat{\mathbf{x}}(t) + \mathbf{B}\mathbf{u}(t) + \mathbf{L}\mathbf{r}(t) \\
\hat{\mathbf{y}}(t) &= \mathbf{C}\hat{\mathbf{x}}(t) + \mathbf{D}\mathbf{u}(t) \\
\mathbf{r}(t) &= \mathbf{y}(t) - \hat{\mathbf{y}}(t)
\end{aligned} \tag{2.11}$$

where $\hat{\mathbf{x}}(t)$ and $\hat{\mathbf{y}}(t)$ represent the estimation of the state and output vectors. $\mathbf{L}$ is the observer gain which is to be designed such that (2.11) is stable and residual signal $\mathbf{r}(t)$ satisfies

$$\forall \mathbf{u}(t), \mathbf{x}(0), \lim_{t \to \infty} \mathbf{r}(t) = 0. \tag{2.12}$$

It is noted that (2.11) is the most widely used residual generator, which is also called fault detection filter (FDF).

To gain a deeper insight into residual generation, the parametrization form of LTI residual generators is introduced in [30].

Theorem 2.1. *Given an LTI system described by (2.3) and let* $\mathbf{G}(s) = \hat{\mathbf{M}}^{-1}(s)\hat{\mathbf{N}}(s)$ *be the left coprime factorization of* $\mathbf{G}(s)$, *then any LTI residual generator can be parameterized by*

$$\mathbf{r}(s) = \mathbf{R}_f(s) \left(\hat{\mathbf{M}}(s)\mathbf{y}(s) - \hat{\mathbf{N}}(s)\mathbf{u}(s) \right) \tag{2.13}$$

where $\mathbf{R}_f(s) \in \mathcal{RH}_\infty$ *is also called post-filter.*

The proof can be found in [35].

The parameterization of residual generators plays an important role in linking residual generation and evaluation, as well as their optimization in the linear observer-based FD framework [32].

Residual Evaluation

Residual evaluation is a procedure for processing residual signal aiming at extracting information of the fault signal. There are two major strategies for the purpose of residual evaluation. Statistic testing is one of them, which is well established in the framework of statistical methods [11]. The other one is the so-called norm-based residual evaluation [43]. In fact, the norm-based residual evaluation is an extended form of the limit monitoring strategy widely used in practice, where a norm of the residual signal, typically either

- $\mathcal{L}_2$-norm

$$J_2(\mathbf{r}) = \|\mathbf{r}(t)\|_2 = \left(\int_0^\infty \mathbf{r}^T(t)\mathbf{r}(t)dt \right)^{1/2} \tag{2.14}$$

or

- $\mathcal{L}_\infty$-norm

$$J_\infty(\mathbf{r}) = \|\mathbf{r}(t)\|_\infty = \sup_t \left(\mathbf{r}^T(t)\mathbf{r}(t) \right)^{1/2} \tag{2.15}$$

is adopted.

Threshold Setting and Decision Logic

Generally speaking, the threshold is set as the possible maximum influence of $\mathbf{x}(0) = \mathbf{x}_0$ and the (bounded) unknown input vector $\mathbf{d}(t)$ on the fault-free residual vector $\mathbf{r}(t)$. Let

$$J_{\text{th},2} = \sup_{\mathbf{x}_0, \mathbf{d}, \mathbf{f}=0} J_2(\mathbf{r}) \tag{2.16}$$

$$J_{\text{th},\infty} = \sup_{\mathbf{x}_0, \mathbf{d}, \mathbf{f}=0} J_\infty(\mathbf{r}) \tag{2.17}$$

be the associated thresholds for evaluation functions (2.14) and (2.15). Thus, the following detection logic

$$\begin{cases} J_2(\mathbf{r}) > J_{\text{th},2} \implies \text{faulty} \\ J_2(\mathbf{r}) \le J_{\text{th},2} \implies \text{fault-free} \end{cases} \tag{2.18}$$

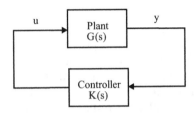

Figure 2.2: Standard feedback control loop

or

$$\begin{cases} J_\infty(\mathbf{r}) > J_{\mathrm{th},\infty} \Longrightarrow \text{faulty} \\ J_\infty(\mathbf{r}) \leq J_{\mathrm{th},\infty} \Longrightarrow \text{fault-free} \end{cases} \tag{2.19}$$

together with residual generator, gives an $\mathcal{L}_2$ or $\mathcal{L}_\infty$ observer-based FD system, respectively.

2.1.3 The Configuration of Observer-based FTC

For this purpose, the stabilizing controllers for LTI systems are characterized in terms of observer-based residual generators.

Observer-based Controller Parametrization Forms

Consider the standard feedback control loop shown in Fig. 2.2, where $\mathbf{G}(s)$ is an LTI system with state space representation (2.1). It is well-known that the Youla-Kucera parametrization

$$\mathbf{K}(s) = \left(\mathbf{Y}(s) - \mathbf{M}(s)\mathbf{Q}_c(s)\right)\left(\mathbf{X}(s) - \mathbf{N}(s)\mathbf{Q}_c(s)\right)^{-1}$$
$$= \left(\hat{\mathbf{X}}(s) - \mathbf{Q}_c(s)\hat{\mathbf{N}}(s)\right)^{-1}\left(\hat{\mathbf{Y}}(s) - \mathbf{Q}_c(s)\hat{\mathbf{M}}(s)\right), \quad \mathbf{Q}_c(s) \in \mathcal{RH}_\infty \tag{2.20}$$

gives all the stabilizing controllers for the plant $\mathbf{G}(s)$ [169]. In [34], an observer-based realization of Youla-Kucera parametrization has been investigated.

Theorem 2.2. *Given the feedback control loop shown in Fig. 2.2 with plant* $\mathbf{G}(s)$ *being factorized by* $\mathbf{G}(s) = \hat{\mathbf{M}}^{-1}(s)\hat{\mathbf{N}}(s) = \mathbf{N}(s)\mathbf{M}^{-1}(s)$, *then any stabilizing controller for* $\mathbf{G}(s)$ *can be parameterized by*

$$\mathbf{u}(s) = \mathbf{F}\hat{\mathbf{x}}(s) + \mathbf{R}(s)\left(\mathbf{y}(s) - \hat{\mathbf{y}}(s)\right)$$
$$\mathbf{R}(s) = -\mathbf{Q}_c(s) \in \mathcal{RH}_\infty. \tag{2.21}$$

Proof. Recall that all proper controllers achieving internal stability for plant $\mathbf{G}(s)$ can be written as

$$\mathbf{K}(s) = \left(\hat{\mathbf{X}}(s) - \mathbf{Q}_c(s)\hat{\mathbf{N}}(s)\right)^{-1}\left(\hat{\mathbf{Y}}(s) - \mathbf{Q}_c(s)\hat{\mathbf{M}}(s)\right) \tag{2.22}$$

which leads to

$$\left(\hat{\mathbf{X}}(s) - \mathbf{Q}_c(s)\hat{\mathbf{N}}(s)\right)\mathbf{u}(s) = \left(\hat{\mathbf{Y}}(s) - \mathbf{Q}_c(s)\hat{\mathbf{M}}(s)\right)\mathbf{y}(s)$$
$$\Rightarrow \hat{\mathbf{X}}(s)\mathbf{u}(s) = \hat{\mathbf{Y}}(s)\mathbf{y}(s) - \mathbf{Q}_c(s)\left(\hat{\mathbf{M}}(s)\mathbf{y}(s) - \hat{\mathbf{N}}(s)\mathbf{u}(s)\right). \tag{2.23}$$

Letting

$$\mathbf{A_L} := \mathbf{A} - \mathbf{LC}, \mathbf{B_L} := \mathbf{B} - \mathbf{LD} \tag{2.24}$$

it follows from the residual generator (2.11) that

$$\hat{\mathbf{x}}(s) = (s\mathbf{I} - \mathbf{A_L})^{-1}(\mathbf{B_L}\mathbf{u}(s) + \mathbf{L}\mathbf{y}(s))$$
$$\mathbf{r}(s) = \hat{\mathbf{M}}(s)\mathbf{y}(s) - \hat{\mathbf{N}}(s)\mathbf{u}(s). \tag{2.25}$$

Furthermore, from

$$\hat{\mathbf{X}}(s) = \mathbf{I} - \mathbf{F}(s\mathbf{I} - \mathbf{A_L})^{-1}\mathbf{B_L}, \hat{\mathbf{Y}}(s) = \mathbf{F}(s\mathbf{I} - \mathbf{A_L})^{-1}\mathbf{L} \tag{2.26}$$

we have that

$$\mathbf{u}(s) = \mathbf{F}(s\mathbf{I} - \mathbf{A_L})(\mathbf{B_L}\mathbf{u}(s) + \mathbf{L}\mathbf{y}(s))$$
$$\quad - \mathbf{Q}_c(s)\left(\hat{\mathbf{M}}(s)\mathbf{y}(s) - \hat{\mathbf{N}}(s)\mathbf{u}(s)\right)$$
$$= \mathbf{F}\hat{\mathbf{x}}(s) + \mathbf{R}(s)\mathbf{r}(s) \tag{2.27}$$

with $\mathbf{R}(s) = -\mathbf{Q}_c(s)$, which completes the proof. $\qquad\square$

Moreover, the following theorem is put forward to reveal a new interpretation of the control loop stabilization and the unique role of the residual signal [31, 153, 165].

Theorem 2.3. *Given the feedback control loop (in Fig. 2.2) with a stabilizing controller $\mathbf{u}_o(s)$, then the set of all controllers that stabilize the given plant $\mathbf{G}(s)$ can be parameterized by*

$$\mathbf{u}(s) = \mathbf{u}_0(s) + \mathbf{R}_c(s)\mathbf{r}(s), \ \mathbf{R}_c(s) \in \mathcal{RH}_\infty \qquad (2.28)$$

Proof. Splitting $\mathbf{R}(s)$ into

$$\mathbf{R}(s) = \mathbf{R}_0(s) + \mathbf{R}_c(s), \ \mathbf{R}_0(s) \in \mathcal{RH}_\infty \qquad (2.29)$$

and substituting into (2.21) gives

$$\mathbf{u}(s) = \mathbf{F}\hat{\mathbf{x}}(s) + \mathbf{R}_0(s)\mathbf{r}(s) + \mathbf{R}_c(s)\mathbf{r}(s), \ \mathbf{R}_c(s) \in \mathcal{RH}_\infty. \qquad (2.30)$$

Noting that

$$\mathbf{u}_0(s) = \mathbf{F}\hat{\mathbf{x}}(s) + \mathbf{R}_0(s)\mathbf{r}(s) \qquad (2.31)$$

spans the space of stabilizing controllers by varying $\mathbf{R}_0(s)$ over all proper, stable, rational matrices. The set of all controllers for plant $\mathbf{G}(s)$ can be parameterized by (2.28), which completes the proof of the theorem. $\square$

Note that the residual signal delivers the information about the unknown inputs in the system. This fact establishes the relationship between control and diagnostics, and motivates the integration of the residual access into the design of FTC systems.

FTC Architecture

Similar to the well-known Youla-Kucera parameterization, Theorem 2.2 and 2.3 also characterize all possible stabilizing controllers for a given plant. The latter term $\mathbf{R}(s)\mathbf{r}(s)$ or $\mathbf{R}_c(s)\mathbf{r}(s)$ denotes the dynamic compensator, and makes no difference on system stability. Thus, it can be employed as a compensatory term to take care of the action caused by disturbing variables or fault variable. Once a stabilizing controller $\mathbf{u}_0(s)$ has been provided, there is a one-to-one correspondence between the parameter matrix $\mathbf{R}(s)$ or $\mathbf{R}_c(s)$ and the controller transfer matrix $\mathbf{K}(s)$.

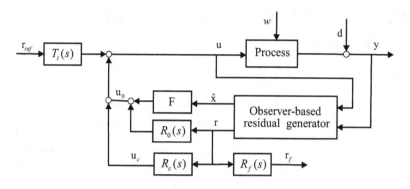

Figure 2.3: The FTC architecture

For FTC purposes, one will be much better off by concentrating on the design of $\mathbf{R}(s)$ or $\mathbf{R}_c(s)$. Also from the viewpoint of practical applications, it would be better to attain performance specifications by only adjusting $\mathbf{R}(s)$ or $\mathbf{R}_c(s)$, instead of redesigning the nominal controller $\mathbf{F}\hat{\mathbf{x}}(s)$ or $\mathbf{u}_0(s)$, which may affect system stability. As a result, together with the parametrization of residual generators (2.13), an observer-based FTC architecture is presented in Fig. 2.3 with

- an observer-based residual generator

$$\dot{\hat{\mathbf{x}}}(t) = \mathbf{A}\hat{\mathbf{x}}(t) + \mathbf{B}\mathbf{u}(t) + \mathbf{L}\mathbf{r}(t)$$
$$\mathbf{r}(t) = \mathbf{y}(t) - \mathbf{C}\hat{\mathbf{x}}(t) - \mathbf{D}\mathbf{u}(t) \qquad (2.32)$$

- a control law

$$\mathbf{u}(s) = \mathbf{u}_0(s) + \mathbf{R}_c(s)\mathbf{r}(s) + \mathbf{T}_t(s)\mathbf{r}_{\text{ref}} \qquad (2.33)$$

- a fault diagnosis algorithm

$$\mathbf{r}_f(s) = \mathbf{R}_f(s)\mathbf{r}(s). \qquad (2.34)$$

$\mathbf{r}_{\text{ref}}$ denotes the reference signal and $\mathbf{T}_t(s)$ is designed to achieve the desired tracking performance. It is noted that for $\mathbf{R}_0(s) = \mathbf{0}, \mathbf{R}_c(s) = \mathbf{0}$, we have a standard observer-based state feedback control. Moreover, $\mathbf{R}_f(s)$ is a post-filter applied to optimizing the FD performance [30].

2.2 FDI and FTC Schemes for Nonlinear Systems

Due to the great effort in the past two decades, observer-based FDI and FTC technique has established itself as an attractive research area in control engineering and theory. Among the actual research topics in this area, nonlinear observer-based FDI and FTC belong definitely to the most challenging issues. In this section, a brief overview on the most widely used observer-based FDI and FTC approaches for nonlinear systems is given. For this purpose, the state space characterizations of different types of nonlinear systems that have been extensively investigated over the past three decades are introduced first.

2.2.1 Representation of Nonlinear Systems

Without loss of generality, nonlinear systems can be modelled as

$$\begin{cases} \dot{\mathbf{x}}(t) = \mathbf{f}(\mathbf{x}(t), \mathbf{u}(t)) \\ \mathbf{y}(t) = \mathbf{h}(\mathbf{x}(t), \mathbf{u}(t)) \end{cases} \tag{2.35}$$

where $\mathbf{x}(t) \in \mathcal{R}^{k_x}, \mathbf{y}(t) \in \mathcal{R}^{k_y}, \mathbf{u}(t) \in \mathcal{R}^{k_u}$ denote the state, output and input vectors, respectively. $\mathbf{f}(\mathbf{x}(t), \mathbf{u}(t))$ and $\mathbf{h}(\mathbf{x}(t), \mathbf{u}(t))$ are continuously differentiable nonlinear function matrices with appropriate dimensions.

To the best of our knowledge, there are few results on FD and FTC studies for the above general type of nonlinear systems since it is difficult to handle the nonlinear dynamics in general. The recent research on nonlinear FD and FTC is mainly dedicated to some special classes of nonlinear systems like Lipschitz nonlinear systems [110, 163], sector bounded nonlinear systems [55], or affine nonlinear systems [1, 73, 72, 151].

Representation of Lipschitz Nonlinear Systems

Without loss of generality, the Lipschitz nonlinear systems can be represented by

$$\begin{cases} \dot{\mathbf{x}}(t) = \mathbf{A}\mathbf{x}(t) + \Gamma(\mathbf{u}, t) + \Phi(\mathbf{x}, \mathbf{u}, t) \\ \mathbf{y}(t) = \mathbf{C}\mathbf{x}(t) + \mathbf{D}\mathbf{u}(t) \end{cases} \tag{2.36}$$

where $\mathbf{A}, \mathbf{C}$ and $\mathbf{D}$ are system matrices with appropriate dimensions; $\Gamma(\mathbf{u}, t)$ is nonlinear function matrix. $\Phi(\mathbf{x}, \mathbf{u}, t)$ is Lipschitz nonlinearity [110, 163] if there exists a positive constant γ, such that for $\forall \mathbf{u} \in \mathcal{R}^{k_u}$ and $t \in \mathcal{R}^+$ and $\forall \mathbf{x}_1, \mathbf{x}_2 \in \mathcal{X}$,

$$||\Phi(\mathbf{x}_1, \mathbf{u}, t) - \Phi(\mathbf{x}_2, \mathbf{u}, t)|| \leq \gamma ||\mathbf{x}_1 - \mathbf{x}_2|| \tag{2.37}$$

where $\mathcal{X}$ is a closed and bounded region containing the origin.

Representation of Sector Bounded Nonlinear Systems

The sector bounded nonlinear systems can be modelled in the following form

$$\begin{cases} \dot{\mathbf{x}}(t) = \mathbf{A}\mathbf{x}(t) + \mathbf{B}\mathbf{u}(t) + \Psi(\mathbf{x}) \\ \mathbf{y}(t) = \mathbf{C}\mathbf{x}(t) + \mathbf{D}\mathbf{u}(t) \end{cases} \tag{2.38}$$

where $\mathbf{A}, \mathbf{B}, \mathbf{C}$ and $\mathbf{D}$ are system matrices with appropriate dimensions. $\Psi(\mathbf{x})$ is a nonlinear function vector satisfying the following sector-bounded condition [55]

$$(\Psi(\mathbf{x}(t)) - \mathbf{T}_1\mathbf{x}(t))^T (\Psi(\mathbf{x}(t)) - \mathbf{T}_2\mathbf{x}(t)) \leq 0 \tag{2.39}$$

for $\mathbf{x}(t) \in \mathcal{R}^{k_x}$. Here, $\mathbf{T}_1$ and $\mathbf{T}_2$ are known real constant matrices and $\mathbf{T} = \mathbf{T}_1 - \mathbf{T}_2$ is symmetric positive definite matrix.

Representation of Affine Nonlinear Systems

The affine nonlinear systems can be described by

$$\begin{cases} \dot{\mathbf{x}}(t) = \mathbf{a}(\mathbf{x}) + \mathbf{b}(\mathbf{x})\mathbf{u}(t) \\ \mathbf{y}(t) = \mathbf{c}(\mathbf{x}) + \mathbf{d}(\mathbf{x})\mathbf{u}(t) \end{cases} \tag{2.40}$$

with $\mathbf{a}(\mathbf{x}), \mathbf{b}(\mathbf{x}), \mathbf{c}(\mathbf{x})$ and $\mathbf{d}(\mathbf{x})$ being continuously differentiable and of appropriate dimensions. This class of nonlinear systems can be considered as a natural extension of LTI systems (2.1).

2.2.2 Classification of Observer-based FDI Techniques

A review of the literature over the past two decades shows that the application of nonlinear observer theory built the main stream in the nonlinear observer-based FDI study in the 90's [2]. In recent years, much attention has been paid to the application of those techniques to addressing nonlinear FDI issues, which are newly established for dealing with analysis and synthesis of nonlinear dynamic systems more efficiently. For instance, adaptive observer-based FDI, sliding mode observer-based FD, Linear parameter-varying (LPV)-based FDI, geometric approach based FDI, artificial intelligence-based approach or fuzzy technique-based FDI have been reported.

Adaptive Observer-based Approach

Generally speaking, adaptive observers adopt recursive algorithms for joint estimation of states and parameters [161]. Due to its constructive nature and the global convergence ensured by an easy-to-check persistent excitation condition, it has been widely investigated for nonlinear FD purpose [145, 163, 140, 17, 36, 51, 61], in particular when the parameters are directly or indirectly related to fault variables. For instance, [61] has addressed the FD problem for a class of nonlinear systems with uncertainties by transforming the systems into two different subsystems under some geometric conditions. Furthermore, [145] has proposed an adaptive fault diagnosis scheme by nontrivially combining a high gain observer and a linear adaptive observer, which is applicable to a class of systems which are truly nonlinear in the sense that they cannot be linearized by coordinate change and output injection. [163] has presented an FDI approach for a class of Lipschitz nonlinear systems with nonlinearity and unstructured modeling uncertainty by adopting an adaptive FD estimator and a bank of fault isolation estimator. In general, the adaptive observer-based FDI approach is more applicable to nonlinear systems with linear parametric uncertainties or faults.

Sliding Mode Observer-based Approach

Sliding mode observers have good robustness against system uncertainties and external disturbances [39, 127], which provides potential for solving robust FDI and FTC problems of nonlinear systems

[62, 42, 149, 148, 27, 147, 3, 86]. In [62], a robust fault estimation scheme has been proposed for affine nonlinear systems in a geometric context. [42] has presented a sliding mode disturbance observer to deal with robust fault detection problem for the affine nonlinear systems. However, both of these two schemes require strict geometric conditions, thus their applicability in practice is restricted. In [147], a robust fault estimation and reconstruction approach for a class of nonlinear systems with uncertainties has been proposed with the aid of convex optimization technique. It has been proved that the reconstructed signal can approximate the fault signal to any accuracy even in the presence of uncertainties. Moreover, the application to a class of nonlinear large-scale systems and civil aircrafts has been reported in [148, 3]. To sum up, sliding mode observer-based FD approach has been widely applied to systems with bounded or Lipschitz nonlinear terms.

LPV-based Approach

LPV techniques have received considerable attention and been intensively investigated for vehicle and aerospace control [15]. It can be considered as an extension of the gain-scheduled control for nonlinear systems. However, this approach is mainly dedicated to nonlinear plants that can be treated as linear systems with on-line measurable and time-varying parameters [15]. More design schemes of integrated fault detection and control problem for LPV systems and FDI filter for LPV systems under a sensitivity constraint can be found in the literatures [7, 141].

Geometric Approach

Geometric approach was first proposed in [97, 98, 79] to solve the design problems of FDF for linear systems. The basic idea behind these approaches is to reconstruct a subsystem which is decoupled from disturbance and only affected by faults. A differential-geometric method has been presented in [109] to address FDI issues for affine nonlinear systems. A basic existence condition of the changes of coordinates in the output space and in the state space has been derived under a mild hypothesis. In [38], the development of residual generators of a class of input affine nonlinear systems has been studied for the purpose of fault reconstruction by means of dynamic system inversion. However, this requires strict geometric conditions which limit its application in practice.

Artificial Intelligence-based Approach

Associated with the rapid development of computer techniques, artificial intelligence-based FD approach has attracted considerable attention in recent years [68, 76]. It has been recognized to be a promising tool for solving FD issues for nonlinear systems when no accurate mathematic model is provided. Signal processing- and knowledge-based approaches serve as the major methodologies. Among the approaches, neural network-based FD methods have been widely investigated due to their adaptation ability to practical processes. However, further efforts are needed for determining the structure and scale of the network, improving the convergence rate and real-time ability and guaranteeing the integrity of training examples.

Fuzzy Technique-based Approach

Over the past decades, the application of Takagi-Sugeno (T-S) fuzzy-model-based analysis and synthesis techniques to deal with nonlinear systems or even non-analytic systems have received intensive attention from both research and application fields [77, 40]. It has been demonstrated that by means of fuzzy dynamic modelling technique, the controller/filtering design issues for nonlinear systems can be transformed into solving a class of linear matrix inequalities (LMIs) in the Lyapunov-function-based framework [128, 60, 48, 155, 100, 21, 23]. Under such circumstances, fault detection problems have been studied in [49, 102, 37, 166, 144, 20, 167]. Among involved studies, [167] has proposed a parity-equation FD approach and a fuzzy-observer-based FD approach for NCSs represented by T-S fuzzy model with different network-induced delays. [102] has proposed a multi-objective FD filter for uncertain T-S fuzzy models so that the residual signal is sensitive to fault signal and meanwhile robust to exogenous input. In [49], a novel fuzzy observer-based design approach is developed to achieve a sensor fault estimation for T-S fuzzy systems with unknown output disturbances. [144] has presented two robust FD schemes (fuzzy-rule-independent and fuzzy-rule-dependent) for T-S fuzzy Itô stochastic systems such that a prescribed noise attenuation level is guaranteed in the $\mathcal{H}_\infty$ sense. In [37], a robust FD approach is developed for a class of uncertain discrete-time T-S fuzzy systems in NCSs with mixed time delays and successive packet dropouts. In [20], an $\mathcal{H}_-/\mathcal{H}_\infty$ FD filter has been proposed for nonlinear systems described by T-S fuzzy models with sensor faults and unknown bounded disturbances by using descriptor approach and non-quadratic Lyapunov functions. Some work on the FD issues for fuzzy systems in presence of unmeasurable premise variables is reported in [6].

In fact, the basic idea behind these approaches lies in lumping the design schemes into finding a common Lyapunov function of a set of LMIs for all subsystems. The conservatism inherent in these common Lypuanov-function-based approaches has motivated investigations on the design methods by means of piecewise/fuzzy Lyapunov functions [166, 159, 41, 66, 157, 111, 112, 114]. The FD problem for fuzzy systems with intermittent measurements has been investigated using basis-dependent Lyapunov functions in [166]. More significantly, both robust full-order fault estimation observer and reduced-order fault estimation observer have been studied for discrete-time T-S fuzzy systems based on piecewise Lyapunov functions in [159]. Roughly speaking, most of these studies have been mainly dedicated to FD design issues for affine systems.

2.2.3 Classification of FTC Techniques for Nonlinear Systems

In the following, the state-of-the-art of FTC techniques for nonlinear systems is reviewed.

Reliable FTC Approach

The basic idea behind a reliable FTC approach lies in taking into account the prior knowledge of potential faults in the controller design. Significant effort has been devoted to study the FTC design for affine nonlinear systems or a class of nonlinear systems with norm-bounded uncertainties in terms of solutions of Hamilton-Jacobi inequalities (HJIs) [85, 90, 150, 115]. Moreover, [13] introduced a passive FTC scheme for affine nonlinear systems with actuator faults. The Lyapunov-based feedback controllers have been addressed to ensure the local uniform asymptotic stability of the systems under additive and loss-of-effectiveness faults. The drawback of these approaches is that the solvability of HJIs is difficult to evaluate.

Adaptive FTC Approach

In order to limit the restrictions on process model accuracy, adaptive techniques have been applied to estimate states and parameters. Based on state and parameter estimation, fault-tolerant performance can be achieved. In [162], a unified approach for FDI and fault accommodation has been proposed for a type of multivariable nonlinear systems by invoking a neural network-based adaptive technique. In [64], a fault accommodation scheme is developed for Lipschitz nonlinear systems with an embedded adaptive fault estimation module. The inherent drawback of an adaptive observer limits its application to more general types of nonlinear systems.

Sliding Mode Based FTC Approach

It is known that sliding mode control has good robustness against uncertainties and external disturbances. In [87], a novel proportional and derivative (PD) sliding mode observer is proposed for nonlinear Markovian jump systems with time delay to simultaneously estimate the state and fault variables. Based on this estimation, an observer-based FTC scheme is developed to guarantee the stochastic stability of the closed-loop system. Furthermore, the application of integral sliding mode FTC approach for the longitudinal control of an aircraft can be found in [4]. For detailed discussions on FTC using sliding mode observers, the readers can refer to [5]. Similar to adaptive observer-based FD approaches, the application of adaptive FTC methods are mainly restricted to some special types of nonlinear systems.

Artificial Intelligence-based FTC Approaches

In recent years, the application of artificial intelligence techniques to deal with nonlinear FTC issues has received significant attention [29, 122]. Generally speaking, this FTC approach is realized by following two steps: (i) adopting the neural network-based learning method to approximate fault models or weighting factors of faults for nonlinear systems, and (ii) re-configuring the control law based on the parameter changes. However, there are few results on the systematic design of neural network, which limits the application of artificial intelligence in solving practical control problems.

Fuzzy Technique-based FTC Approach

As mentioned earlier, T-S fuzzy-model-based analysis and synthesis technique has been well recognized as a simple and effective tool to controlling nonlinear systems or even nonanalytic systems [126, 156, 18]. More recently, there have also appeared some results on fuzzy-model-based FTC for nonlinear systems [59, 132]. It has been demonstrated that by using the fuzzy dynamic modelling technique, the FTC design issues for nonlinear systems can be transformed to solving a class of LMIs in the Lyapunov-function-based framework. This inherent advantage motivates its application in real industrial processes [70, 28, 120].

2.3 Concluding Remarks

This chapter represents FD and FTC techniques for both linear and nonlinear systems. A brief description on the developments of FD and FTC configurations for linear systems has been provided. It is worth mentioning that the parametrization forms of residual generators and controllers play an essential role in obtaining the associated configurations. However, the generalization of the parametrization forms to nonlinear systems seems to be left out of investigation. Therefore, the configurations of observer-based FD and FTC for nonlinear systems will be studied in Chapters 3 and 7. Then, different types of nonlinear processes have been introduced. The major FD and FTC methodologies have been recalled for nonlinear systems. It can be observed that only some of these studies have dealt with residual generator and evaluation as well as decision making in an integrated way. Most of efforts have been made on the FD system design but little on the analysis issues. Thus, the forthcoming chapter discusses the analysis and integrated design issues for nonlinear observer-based FD systems. The integrated design of observer-based FD systems is now basically a blank field for the research of general types of nonlinear processes. To deal with the nonlinear issues, a mathematical and systematic tool is needed. Inspirited by the T-S-Fuzzy controller design for nonlinear systems, the investigations on fuzzy observer-based FD systems for general types of nonlinear systems are addressed in Chapters 4-5.

3 Configuration of Nonlinear Observer-based FD Systems

As discussed in Chapter 2, an observer-based FD system consists of an observer-based residual generator, a residual evaluator and a decision maker with an embedded threshold. In the past two decades, great efforts have been made to the development of observer-based FD schemes for nonlinear systems, where the major focus is on the design of nonlinear observer-based residual generators, as summarized in the review papers [2, 16]. It has been observed that only few of recent studies have dealt with residual generator and evaluation as well as decision making in an integrated way [1, 73, 72]. In addition, most of research efforts on the FD design issues have been addressed based on an implicit assumption that a stable residual generator can be constructed. However, there is no commonly used condition for checking the existence of an observer-based FD system for general type of nonlinear systems. On the other hand, in the linear observer-based FDI framework, parametrization of residual generators plays an important role in linking the residual generation and evaluation and their optimization [30]. In fact, a parametrization form of nonlinear observer-based residual generators has been applied to the study on optimizing residual evaluation and threshold setting [1]. Besides, there have been very few results presented on the systematic derivation of the parameterization form.

Motivated by these observations, in the first part of this chapter an essential analysis issue on the existence conditions of nonlinear observer-based FD systems is addressed. With the aid of input-to-output (IOS) theory, the concepts $\mathcal{L}_\infty$ re-constructability, $\mathcal{L}_2$ re-constructability and $\mathcal{L}_\infty/\mathcal{L}_2$ re-constructability are introduced and, in this context, existence conditions for different types of nonlinear observer-based FD systems are investigated. The objective of this work is to gain a deeper insight into the fundamental properties of nonlinear observer-based FD systems, which may be useful for the development of nonlinear FD systems using some well established technologies. In the second part of this chapter, the parameterization issues for nonlinear observer-based FD systems are addressed. The parameterize form of the observer-based residual generators is first presented by means of the nonlinear factorization and input-output operator techniques, which, similar to the LTI case, is composed of a kernel system (of the system under consideration) and a post-filter. Based on a state-space realization of the proposed parametrization, a characterization of the overall observer-based FD systems including the residual evaluator and the threshold setting will then be studied. To this end, the concept of the IOS and some methods for system input/output stabilization will be applied.

3.1 Preliminaries and Problem Formulation

The nonlinear systems under consideration can be described by

$$\Sigma^{\mathbf{x}_0} : \begin{cases} \dot{\mathbf{x}} = \mathbf{f}(\mathbf{x}, \mathbf{u}) \\ \mathbf{y} = \mathbf{h}(\mathbf{x}, \mathbf{u}) \end{cases} \tag{3.1}$$

where $\mathbf{x} \in \mathcal{R}^{k_x}, \mathbf{u} \in \mathcal{R}^{k_u}, \mathbf{y} \in \mathcal{R}^{k_y}$ denote the state, output and input vectors, respectively. $\mathbf{f}(\mathbf{x}, \mathbf{u})$ and $\mathbf{h}(\mathbf{x}, \mathbf{u})$ are continuously differentiable nonlinear functions with appropriate dimensions. System (3.1) is called faulty if undesirable changes in the system dynamics are caused by some faults. The faulty system dynamics is often modelled by

$$\Sigma_f^{\mathbf{x}_0} : \begin{cases} \dot{\mathbf{x}} = \tilde{\mathbf{f}}(\mathbf{x}, \mathbf{u}, \mathbf{w}) \\ \mathbf{y} = \tilde{\mathbf{h}}(\mathbf{x}, \mathbf{u}, \mathbf{w}) \end{cases} \tag{3.2}$$

with $\mathbf{w} \in \mathcal{R}^{k_w}$ denoting the fault vector. It is noted that $\tilde{\mathbf{f}}(\mathbf{x}, \mathbf{u}, \mathbf{0}) = \mathbf{f}(\mathbf{x}, \mathbf{u}), \tilde{\mathbf{h}}(\mathbf{x}, \mathbf{u}, \mathbf{0}) = \mathbf{h}(\mathbf{x}, \mathbf{u})$.

For FD purpose, an observer-based FD system for nonlinear processes is typically configured with an observer-based residual generator, an evaluation function and a decision maker with a threshold [151]. The definition for nonlinear residual generators is first introduced as follows.

Definition 3.1. *Given nonlinear system (3.1), a system of the form*

$$\begin{cases} \dot{\hat{\mathbf{x}}} = \phi(\hat{\mathbf{x}}, \mathbf{u}, \mathbf{y}) \\ \mathbf{r} = \varphi(\hat{\mathbf{x}}, \mathbf{u}, \mathbf{y}) \end{cases} \tag{3.3}$$

is called observer-based residual generator if it delivers a residual vector
$\mathbf{r}$ *satisfying that (i) for* $\hat{\mathbf{x}}(0) = \mathbf{x}(0), \forall \mathbf{u}(t), \mathbf{r}(t) \equiv \mathbf{0}$ *(ii) for some* $\mathbf{w} \neq \mathbf{0}$
in the faulty system (3.2), $\mathbf{r}(t) \neq \mathbf{0}$.

Residual evaluation is a procedure for residual signal processing aiming at creating a mathematical feature of the residual signal, i.e., $J(\mathbf{r})$. From the engineering viewpoint, the determination of a threshold J_{th} is equivalent to find out the tolerant limit for unknown/known inputs and initial conditions on residual signal in fault-free case, i.e.,

$$J_{\text{th}} = \sup_{\mathbf{w}=\mathbf{0}, \mathbf{u}, \mathbf{x}(0), \hat{\mathbf{x}}(0)} J(\mathbf{r}). \tag{3.4}$$

Then, the detection logic

$$\begin{cases} J(\mathbf{r}) > J_{\text{th}} \implies \text{faulty} \\ J(\mathbf{r}) \leq J_{\text{th}} \implies \text{fault-free} \end{cases} \tag{3.5}$$

can be finally used for decision making.

The problems to be addressed in this chapter can be formulated in the following three parts:

- The definitions and the existence conditions of observer-based FD systems for nonlinear systems (3.1) are studied.

- The parameterization form of nonlinear observer-based residual generators is addressed by means of the nonlinear factorization and input-output operator techniques.

- The parameterized residual generators are characterized in the state space representation. And the parameterization of the overall observer-based FD systems will be achieved by applying the concept of the IOS and methods for system input/output stabilization.

3.2 On the Existence Conditions of Observer-based FD Systems

To avoid loss of information about the faults, the residual vector has generally the same dimension like the output vector. For the sake of simplicity and also considering the conditions (i) and (ii) given in Definition 3.1, it is supposed that

$$\mathbf{r} = \varphi(\hat{\mathbf{x}}, \mathbf{u}, \mathbf{y}) = \mathbf{y} - \hat{\mathbf{y}}, \hat{\mathbf{y}} = \mathbf{h}(\hat{\mathbf{x}}, \mathbf{u}). \tag{3.6}$$

Thus, in this section, the following type of residual generators is considered

$$\begin{cases} \dot{\hat{\mathbf{x}}} = \phi(\hat{\mathbf{x}}, \mathbf{u}, \mathbf{y}), \hat{\mathbf{y}} = \mathbf{h}(\hat{\mathbf{x}}, \mathbf{u}) \\ \mathbf{r} = \mathbf{y} - \hat{\mathbf{y}}. \end{cases} \tag{3.7}$$

For the residual evaluation purpose, two norm-based evaluation functions are considered in this chapter: (i) the Euclidean norm based instant evaluation

$$J_E(\mathbf{r}) = \varphi_1 \left(\|\mathbf{r}\| \right) \tag{3.8}$$

(ii) integral evaluation with a finite-time evaluation window $[0, \tau]$

$$J_2(\mathbf{r}) = \int_0^\tau \varphi_2 \left(\|\mathbf{r}\| \right) dt \tag{3.9}$$

where $\varphi_1 \left(\|\mathbf{r}\| \right), \varphi_2 \left(\|\mathbf{r}\| \right)$ are some $\mathcal{K}$-functions.

For our purpose, we first introduce the definitions for different types of observer-based FD systems.

Definition 3.2. *Given the nonlinear system (3.1), a dynamic system is called*

- $\mathcal{L}_\infty$ *observer-based FD system, when it consists of the observer-based residual generator (3.3), residual evaluation function (3.8) and detection logic (3.5) with a corresponding threshold*

- $\mathcal{L}_2$ *observer-based FD system, when it consists of the observer-based residual generator (3.3), residual evaluation function (3.9) and detection logic (3.5) with a corresponding threshold.*

- $\mathcal{L}_\infty/\mathcal{L}_2$ *observer-based FD system, when it consists of the observer-based residual generator (3.3), residual evaluation function (3.8) and detection logic (3.5) with a corresponding dynamics threshold.*

In the subsequent three subsections, the existence conditions of the above three types of FD systems as well as the construction of the corresponding thresholds will be studied.

3.2.1 $\mathcal{L}_\infty$ Observer-based FD Systems

For our purpose, the following definition is first introduced, which is motivated by the so-called *weak detectability* given in [138]. The concept weak detectability is widely used in the study on the stabilization of nonlinear systems by output feedback [107, 91].

Definition 3.3. *System (3.1) is said to be $\mathcal{L}_\infty$ re-constructible if there exist (i) a function* $\phi : \mathcal{R}^{k_x} \times \mathcal{R}^{k_u} \times \mathcal{R}^{k_y} \to \mathcal{R}^{k_x}$ *(ii) functions* $V(\mathbf{x}, \hat{\mathbf{x}})$: $\mathcal{R}^{k_x} \times \mathcal{R}^{k_x} \to \mathcal{R}^+, \varphi_i\,(\cdot) \in \mathcal{K}, i = 1, 2, 3,$ *and positive constants* δ, δ_u *such that* $\forall \mathbf{x}, \hat{\mathbf{x}} \in \mathcal{B}_\delta, \|\mathbf{u}\|_\infty \leq \delta_u,$ *the following inequalities hold*

$$\varphi_1\,(\|\mathbf{r}\|) \leq V(\mathbf{x}, \hat{\mathbf{x}}) \leq \varphi_2\,(\|\mathbf{x} - \hat{\mathbf{x}}\|), \mathbf{r} = \mathbf{y} - \mathbf{h}(\hat{\mathbf{x}}, \mathbf{u}) \tag{3.10}$$

$$V_\mathbf{x}(\mathbf{x}, \hat{\mathbf{x}})\mathbf{f}(\mathbf{x}, \mathbf{u}) + V_{\hat{\mathbf{x}}}(\mathbf{x}, \hat{\mathbf{x}})\phi(\hat{\mathbf{x}}, \mathbf{u}, \mathbf{y}) \leq -\varphi_3\,(\|\mathbf{x} - \hat{\mathbf{x}}\|). \tag{3.11}$$

Remark 3.1. *Substituting* $\|\mathbf{r}\|$ *in* $\varphi_1\,(\|\mathbf{r}\|)$ *by* $\|\mathbf{x} - \hat{\mathbf{x}}\|$, *Definition 3.3 becomes equivalent with the weak detectability given in [138]. If it is further assumed, as in [45], that*

$$\|\mathbf{h}(\zeta, \mathbf{u}) - \mathbf{h}(\sigma, \mathbf{u})\| \leq \gamma\,(\|\zeta - \sigma\|) \tag{3.12}$$

for some $\gamma \in \mathcal{K}$, *we have* $\|\mathbf{x} - \hat{\mathbf{x}}\| \geq \gamma^{-1}\,(\|\mathbf{r}\|)$, *which leads to* $\varphi_1\,(\|\mathbf{x} - \hat{\mathbf{x}}\|) \geq \varphi_1\,(\gamma^{-1}\,(\|\mathbf{r}\|))$. *Since* $\varphi_1\,(\gamma^{-1}\,(\cdot)) \in \mathcal{K}$, *the weak detectability implies the output re-constructability.*

Remark 3.2. *Consider the (overall) system*

$$\dot{\mathbf{x}} = \mathbf{f}(\mathbf{x}, \mathbf{u}), \mathbf{y} = \mathbf{h}(\mathbf{x}, \mathbf{u}), \dot{\hat{\mathbf{x}}} = \phi(\hat{\mathbf{x}}, \mathbf{u}, \mathbf{y}), \mathbf{r} = \mathbf{y} - \mathbf{h}(\hat{\mathbf{x}}, \mathbf{u}) \tag{3.13}$$

with $\mathbf{u}$ *as its input and* $\mathbf{r}$ *as output. Function* $V(\mathbf{x}, \hat{\mathbf{x}})$ *satisfying (3.10)-(3.11) can be understood as a variant of the so-called IOS-Lyapunov function [124]. In fact, the residual generation problem can also be studied in the IOS context.*

The following theorem presents a major property of an output re-constructible system, which provides us with a sufficient condition for the existence of an $\mathcal{L}_\infty$ observer-based FD system and the threshold setting.

Theorem 3.1. *Assume that system (3.1) is $\mathcal{L}_\infty$ re-constructible. Then, system (3.7) delivers a residual vector* $\mathbf{r}(t)$, *which satisfies*

$$\|\mathbf{r}(t)\| \leq \beta\,(\|\mathbf{x}(0) - \hat{\mathbf{x}}(0)\|, t) \tag{3.14}$$

where $\beta\,(\|\mathbf{x}(0) - \hat{\mathbf{x}}(0)\|, t) \in \mathcal{KL}$.

Proof. It follows from (3.10) that

$$\|\mathbf{x}(t) - \hat{\mathbf{x}}(t)\| \geq \varphi_2^{-1} \left(V(\mathbf{x}, \hat{\mathbf{x}}) \right) \tag{3.15}$$

by which (3.11) can be further re-written into

$$\dot{V}(\mathbf{x}, \hat{\mathbf{x}}) \leq -\varphi_3 \left(\varphi_2^{-1} \left(V(\mathbf{x}, \hat{\mathbf{x}}) \right) \right). \tag{3.16}$$

Since $\varphi_3(\varphi_2^{-1}) \in \mathcal{K}$, it is known from Lemma 6.1 in [123] that there exists a $\mathcal{KL}$-function $\gamma(\cdot)$ so that

$$V(\mathbf{x}(t), \hat{\mathbf{x}}(t)) \leq \gamma \left(V(\mathbf{x}(0), \hat{\mathbf{x}}(0)), t \right) \tag{3.17}$$

Note that it follows from (3.10) that

$$\|\mathbf{r}(t)\| \leq \varphi_1^{-1} \left(V(\mathbf{x}, \hat{\mathbf{x}}) \right). \tag{3.18}$$

Further considering (3.10) and (3.11), it results in

$$\|\mathbf{r}(t)\| \leq \varphi_1^{-1} \left(\gamma \left(\varphi_2 \left(\|\mathbf{x}(0) - \hat{\mathbf{x}}(0)\| \right), t \right) \right) := \beta \left(\|\mathbf{x}(0) - \hat{\mathbf{x}}(0)\|, t \right). \tag{3.19}$$

The theorem is thus proven. □

An immediate result of Theorem 3.1 is the following corollary, which can be used for the design of an $\mathcal{L}_\infty$ observer-based FD system.

Corollary 3.1. *Assume that system (3.1) is $\mathcal{L}_\infty$ re-constructible, $\|\mathbf{x}(0) - \hat{\mathbf{x}}(0)\| \leq \delta_o$ and the evaluation window is $[t_1, t_2]$, i.e. $J_E = \varphi_1 \left(\|\mathbf{r}(t)\| \right)$, $t \in [t_1, t_2]$. Then, an L_∞ observer-based FD system can be built by (i) constructing residual generator (3.7) using ϕ defined in Definition 3.3; (ii) setting the threshold*

$$J_{\text{th}} = \beta \left(\delta_o, t_1 \right). \tag{3.20}$$

Proof. It follows directly from Theorem 3.1 and the definition of $\mathcal{KL}$-function $\beta \left(\|\mathbf{x}(0) - \hat{\mathbf{x}}(0)\|, t \right)$ that

$$J_{\text{th}} = \max_{\substack{\|\mathbf{x}(0) - \hat{\mathbf{x}}(0)\| \leq \delta_o \\ t \in [t_1, t_2]}} \beta \left(\|\mathbf{x}(0) - \hat{\mathbf{x}}(0)\|, t \right)$$

$$= \beta \left(\max_{\|\mathbf{x}(0) - \hat{\mathbf{x}}(0)\| \leq \delta_o} \|\mathbf{x}(0) - \hat{\mathbf{x}}(0)\|, \min_{t \in [t_1, t_2]} t \right) = \beta \left(\delta_o, t_1 \right). \tag{3.21}$$

The proof is completed. □

Corollary 3.1 reveals that if system (3.1) is $\mathcal{L}_\infty$ re-constructible, a constant threshold can be found. However, such a setting could be too conservative. In order to improve the FD performance, the influence of the process input variables on the residual vector should be generally taken into account. It will lead to a so-called adaptive threshold, which allows a more efficient fault detection than a constant one [30]. For this purpose, the following detection scheme is presented.

3.2.2 $\mathcal{L}_2$ Observer-based FD Systems

Considering that the $\mathcal{L}_2$-norm is often used for the residual evaluation purpose, the following $\mathcal{L}_2$ type of re-constructability is introduced.

Definition 3.4. *System (3.1) is said to be $\mathcal{L}_2$ re-constructible if there exist (i) a function $\phi : \mathcal{R}^n \times \mathcal{R}^p \times \mathcal{R}^m \to \mathcal{R}^n$ (ii) functions $V(\mathbf{x}, \hat{\mathbf{x}})$: $\mathcal{R}^n \times \mathcal{R}^n \to \mathcal{R}^+, \varphi_1(\cdot) \in \mathcal{K}, \varphi_2(\cdot) \in \mathcal{K}_\infty$ and positive constants δ, δ_u such that $\forall \mathbf{x}, \hat{\mathbf{x}} \in \mathcal{B}_\delta, \|\mathbf{u}\|_\infty \leq \delta_u$*

$$V_{\mathbf{x}}(\mathbf{x}, \hat{\mathbf{x}})\mathbf{f}(\mathbf{x}, \mathbf{u}) + V_{\hat{\mathbf{x}}}(\mathbf{x}, \hat{\mathbf{x}})\phi(\hat{\mathbf{x}}, \mathbf{u}, \mathbf{h}(\mathbf{x}, \mathbf{u})) \leq -\varphi_1(\|\mathbf{r}\|) + \varphi_2(\|\mathbf{u}\|). \quad (3.22)$$

Comparing Definitions 3.3 and 3.4, it becomes evident that condition (3.22) is generally weaker than the conditions given in Definition 3.3.

Remark 3.3. *Corresponding to the discussion on IOS and $\mathcal{L}_\infty$ re-constructability in Remark 3.2, function $V(\mathbf{x}, \hat{\mathbf{x}})$ given in the above definition is in fact a variant of the ROS-Lyapunov function (robust output stability) [124].*

The following theorem presents the major result of this subsection, a sufficient condition for the existence of an $\mathcal{L}_2$ observer-based FD system.

Theorem 3.2. *Assume that system (3.1) is $\mathcal{L}_2$ re-constructible. Then, an $\mathcal{L}_2$ observer-based FD system can be realized using functions $\varphi_i, i = 1, 2$ and by (i) constructing residual generator according to (3.7) (ii) defining the residual evaluation function as*

$$J(\mathbf{r}) = \int_0^\tau \varphi_1(\|\mathbf{r}(t)\|)\, dt \quad (3.23)$$

and (iii) setting the threshold

$$J_{\mathrm{th}} = \int_0^\tau \varphi_2(\|\mathbf{u}(t)\|)\, dt + \bar{\gamma}_o, \bar{\gamma}_o = \sup_{\mathbf{x}(0), \hat{\mathbf{x}}(0)} \{\gamma_0\}, \gamma_o = V(\mathbf{x}(0), \hat{\mathbf{x}}(0)).$$
$$(3.24)$$

Proof. It follows from (3.22) that

$$\dot{V}(\mathbf{x}(t), \hat{\mathbf{x}}(t)) \leq -\varphi_1\left(\|\mathbf{r}(t)\|\right) + \varphi_2\left(\|\mathbf{u}(t)\|\right) \tag{3.25}$$

which leads to

$$\int_0^\tau \varphi_1\left(\|\mathbf{r}(t)\|\right) dt + V(\mathbf{x}(\tau), \hat{\mathbf{x}}(\tau)) \leq \int_0^\tau \varphi_2\left(\|\mathbf{u}(t)\|\right)) dt + V(\mathbf{x}(0), \hat{\mathbf{x}}(0)). \tag{3.26}$$

As a result,

$$\int_0^\tau \varphi_1\left(\|\mathbf{r}(t)\|\right)) dt \leq \int_0^\tau \varphi_2\left(\|\mathbf{u}(t)\|\right)) dt + V(\mathbf{x}(0), \hat{\mathbf{x}}(0)). \tag{3.27}$$

The theorem is thus proved. □

In theoretical study on norm-based residual evaluation, it is the state of the art that the evaluation window is assumed to be infinitively large, i.e.

$$J(\mathbf{r}) = \int_0^\infty \varphi_5\left(\|\mathbf{r}(t)\|\right) dt \Longrightarrow J_{\text{th}} = \int_0^\infty \varphi_4\left(\|\mathbf{u}(t)\|\right) dt + \bar{\gamma}_o. \tag{3.28}$$

However, in practice, it is not realistic, since a large evaluation window generally results in a (considerably) delayed fault detection. In dealing with nonlinear FD, a large evaluation window also means a high threshold due to the dependence on $\mathbf{u}$. Considering that a fault may happen after the system is in operation for a long time, with a large evaluation window the influence of $\mathbf{w}$ on $J(\mathbf{r})$ may be much weaker than $\mathbf{u}$ on J_{th}. As a result, the FD performance will become poor. For these reasons, in practice the evaluation function and threshold are often defined by

$$J(\mathbf{r}) = \int_{t_o}^{t_o+\tau} \varphi_5\left(\|\mathbf{r}(t)\|\right) dt \Longrightarrow J_{\text{th}} = \int_{t_o}^{t_o+\tau} \varphi_4\left(\|\mathbf{u}(t)\|\right) dt + \bar{\gamma}_o \tag{3.29}$$

where $\bar{\gamma}_o = \sup\limits_{\mathbf{x}(t_0), \hat{\mathbf{x}}(t_0)} \{\gamma_o\}$ represents the maximum value of γ_o for all (bounded) possible $\mathbf{x}(t_0), \hat{\mathbf{x}}(t_0)$.

3.2.3 $\mathcal{L}_\infty/\mathcal{L}_2$ Observer-based FD Systems

It is noted that $\mathcal{L}_2$-norm measures the "average energy level" in a signal over the time interval $[0, \tau]$ and is suitable to measure the "size" of a slowly varying signal. In against, $\mathcal{L}_\infty$-norm measures the (maximum) "size" of a signal at each time instant and therefore is often used to measure the "size" of a rapidly changing signal. In practice, both the energy level and the maximum absolute value are widely used for evaluating the residual signals. From application perspective, an early detection of faults could avoid or minimize disastrous consequences. To this end, $\mathcal{L}_\infty$-norm is adopted for residual evaluation purpose in this work, which plays an essential rule in realizing a real-time detection of faults that would cause an abrupt change in the residual signal.

It follows from Corollary 3.1 that if there exists a residual generator (3.7) and $\beta(\cdot, \cdot) \in \mathcal{KL}$ such that

$$
\begin{aligned}
\varphi_1\left(\|\mathbf{r}(t)\|\right) &\leq \beta(\|\mathbf{x}(0) - \hat{\mathbf{x}}(0)\|, t) \\
&\leq \max_{t \in [t_1, t_2]} \beta(\|\mathbf{x}(0) - \hat{\mathbf{x}}(0)\|, t) \\
&= \beta(\|\mathbf{x}(0) - \hat{\mathbf{x}}(0)\|, t_1) := J_{\text{th}}
\end{aligned}
\tag{3.30}
$$

then an $\mathcal{L}_\infty$ observer-based FD system can be constructed. However, this approach is very conservative since the influence of input signal on residual signal for nonlinear systems can not be fully compensated and should be taken into consideration in the observer-based FD scheme. This motivated us to investigate the following FD scheme.

Definition 3.5. *System (3.1) is said to be $\mathcal{L}_\infty/\mathcal{L}_2$ re-constructible if there exists a nonlinear system*

$$
\begin{cases}
\dot{\hat{\mathbf{x}}} = \phi(\hat{\mathbf{x}}, \mathbf{u}, \mathbf{y}) \\
\hat{\mathbf{y}} = \mathbf{h}(\hat{\mathbf{x}})
\end{cases}
\tag{3.31}
$$

such that $\forall \mathbf{x}, \hat{\mathbf{x}} \in \mathcal{B}_\delta$

$$
\varphi_1(\|\mathbf{y} - \hat{\mathbf{y}}\|) \leq \int_0^\tau \varphi_2(\|\mathbf{u}(t)\|)dt + \gamma_o(\mathbf{x}(0), \hat{\mathbf{x}}(0))
\tag{3.32}
$$

where $\varphi_1(\cdot) \in \mathcal{K}, \varphi_2(\cdot) \in \mathcal{K}_\infty$, $\delta > 0$ *and* $\gamma_o(\cdot) \geq 0$ *is a (finite) constant for given* $\mathbf{x}(0), \hat{\mathbf{x}}(0)$.

A sufficient condition for the $\mathcal{L}_\infty/\mathcal{L}_2$ type of re-constructability is introduced in the following theorem, which also serves as the existence condition for an $\mathcal{L}_\infty/\mathcal{L}_2$ type of observer-based FD systems and the threshold setting.

Theorem 3.3. *Given system (3.1), if there exist (i) a function* $\phi : \mathcal{R}^{k_x} \times \mathcal{R}^{k_u} \times \mathcal{R}^{k_y} \to \mathcal{R}^{k_x}$; *(ii) functions* $V(\mathbf{x}, \hat{\mathbf{x}}) : \mathcal{R}^{k_x} \times \mathcal{R}^{k_x} \to \mathcal{R}^+, \varphi_1(\cdot) \in \mathcal{K}, \varphi_2(\cdot) \in \mathcal{K}_\infty, \varphi_3(\cdot) \in \mathcal{K}_\infty$ *and a positive constant* δ *such that* $\forall \mathbf{x}, \hat{\mathbf{x}} \in \mathcal{B}_\delta$

$$\varphi_1(\|\mathbf{r}\|) \leq V(\mathbf{x}, \hat{\mathbf{x}}), \quad \mathbf{r} = \mathbf{y} - \hat{\mathbf{y}} \tag{3.33}$$

$$V_{\mathbf{x}}(\mathbf{x}, \hat{\mathbf{x}})\mathbf{f}(\mathbf{x}, \mathbf{u}) + V_{\hat{\mathbf{x}}}(\mathbf{x}, \hat{\mathbf{x}})\phi(\hat{\mathbf{x}}, \mathbf{u}, \mathbf{h}(\mathbf{x}, \mathbf{u})) \leq \varphi_2(\|\mathbf{u}\|) \tag{3.34}$$

then system (3.1) is $\mathcal{L}_\infty/\mathcal{L}_2$ *re-constructible. Moreover, an* $\mathcal{L}_\infty/\mathcal{L}_2$ *type of observer-based FD system can be realized by (i) constructing residual generator (3.7) (ii) defining the residual evaluation function*

$$J_E(\mathbf{r}) = \varphi_1(\|\mathbf{r}(t)\|) \tag{3.35}$$

and (iii) setting the threshold

$$J_{\text{th}} = \int_0^T \varphi_2\left(\|\mathbf{u}(t)\|\right)dt + \bar{\gamma}_0$$

$$\bar{\gamma}_0 = \sup_{\mathbf{x}(0), \hat{\mathbf{x}}(0) \in \mathcal{B}_\delta} V\left(\mathbf{x}(0), \hat{\mathbf{x}}(0)\right). \tag{3.36}$$

Proof. It follows directly from (3.34) that

$$V(\mathbf{x}(\tau), \hat{\mathbf{x}}(\tau)) - V(\mathbf{x}(0), \hat{\mathbf{x}}(0)) \leq \int_0^T \varphi_2\left(\|\mathbf{u}(t)\|\right)dt. \tag{3.37}$$

Then, together with (3.33), we have

$$\varphi_1(\|\mathbf{r}(\tau)\|) \leq \int_0^T \varphi_2\left(\|\mathbf{u}(t)\|\right)dt + V(\mathbf{x}(0), \hat{\mathbf{x}}(0)) \leq \int_0^T \varphi_2\left(\|\mathbf{u}(t)\|\right)dt + \bar{\gamma}_0 \tag{3.38}$$

which completes the proof. $\square$

In this section, the existence conditions for three different types of nonlinear observer-based FD systems have been derived. Although the achieved results do not lead to a direct design of a nonlinear observer-based FD system, they are fundamental for the application of some well established nonlinear techniques for FD system design. For instance, inspirited by the T-S-Fuzzy controller design for nonlinear systems [47, 48], in the forthcoming chapters, some results on fuzzy observer-based FD system design will be presented based on the conditions given in Theorems 3.2-3.3, respectively.

3.3 Parametrization of Nonlinear Residual Generators

In this section, the parametrization of residual generators for system (3.1) will be addressed. This work is mainly based on the input-output operator approach described in [106, 105]. To ease the derivations, the following notations are adopted. A signal space $\mathcal{U}$ denotes a vector space of functions from a time domain to an Euclidean vector space. $\mathcal{U}^s$ represents the stable subset of $\mathcal{U}$. An operator Σ with an input signal space $\mathcal{U}$, an output signal space $\mathcal{Y}$ and an initial condition $x_0 \in \mathcal{X}_0$ is denoted by $\Sigma^{x_0} : \mathcal{U} \to \mathcal{Y}$. It is said to be stable if $\forall \mathbf{x}_0, \mathbf{u} \in \mathcal{U}^s \Rightarrow \Sigma^{\mathbf{x}_0}(\mathbf{u}) \in \mathcal{Y}^s$. The cascade connection of two systems $\Sigma_1^{\xi_0} : \mathcal{U} \times \mathcal{Y} \to \mathcal{Z}$ and $\Sigma_2^{\varsigma_0} : \mathcal{L} \to \mathcal{U} \times \mathcal{Y}$ is denoted by $\Sigma_1^{\xi_0} \circ \Sigma_2^{\varsigma_0} : \mathcal{L} \to \mathcal{Z}$.

Let $\Sigma^{x_0} : \mathcal{U} \to \mathcal{Y}$ and $\Sigma_f^{\mathbf{x}_0} : \mathcal{U} \times \mathcal{W} \to \mathcal{Y}$ be the operators of (3.1) and (3.2) respectively, and assume that $\Sigma^{\mathbf{x}_0}(\mathbf{u}) = \Sigma_f^{\mathbf{x}_0}(\mathbf{u}, \mathbf{0})$.

Definition 3.6. *Given $\Sigma^{\mathbf{x}_0}, \Sigma_f^{\mathbf{x}_0}$, the fault vector $\mathbf{w}(\neq \mathbf{0})$ is said to be detectable if for some $\mathbf{u}, \mathbf{x}_0, \Sigma^{\mathbf{x}_0}(\mathbf{u}) \neq \Sigma_f^{\mathbf{x}_0}(\mathbf{u}, \mathbf{w})$.*

It is reasonable that in this study only detectable faults are considered.

Definition 3.7. *An operator $R_\Sigma^{\xi_0} : \mathcal{U}^s \times \mathcal{Y}^s \to \mathcal{R}^s$ is called (stable) residual generator if*

$$
\begin{cases}
\forall \mathbf{u}, \mathbf{x}_0, \exists \xi_0 \text{ so that } R_\Sigma^{\xi_0}\begin{pmatrix} \mathbf{u} \\ \mathbf{y} \end{pmatrix} = \mathbf{0} \text{ for } \mathbf{w} = \mathbf{0} \\
R_\Sigma^{\xi_0}\begin{pmatrix} \mathbf{u} \\ \mathbf{y} \end{pmatrix} \neq \mathbf{0} \text{ for detectable } \mathbf{w} \neq \mathbf{0}.
\end{cases}
\tag{3.39}
$$

The output of $R_\Sigma^{\xi_0}, \mathbf{r} = R_\Sigma^{\xi_0}\begin{pmatrix} \mathbf{u} \\ \mathbf{y} \end{pmatrix}$, is called residual vector.

Condition (3.39) means that the residual generator is driven by the process input and output vectors $\mathbf{u}, \mathbf{y}$, and the residual vector should be zero in the fault-free case. Note that $\mathbf{x}, \xi$ may have different dimensions.

In what follows, kernel representation and image representation are introduced which are the generalizations of left coprime factorization and right coprime factorization to nonlinear systems.

Definition 3.8. *A kernel representation of an operator* $\Sigma^{\mathbf{x}_0} : \mathcal{U} \to \mathcal{Y}$ *is any operator* $K_{\Sigma}^{\hat{\mathbf{x}}_0} : \mathcal{U} \times \mathcal{Y} \to \mathcal{R}$, *such that for* $\forall \hat{\mathbf{x}}_0 = \mathbf{x}_0$

$$\mathbf{y} = \Sigma^{\mathbf{x}_0}(\mathbf{u}) \implies K_{\Sigma}^{\hat{\mathbf{x}}_0} \begin{pmatrix} \mathbf{u} \\ \mathbf{y} \end{pmatrix} = \mathbf{0}. \tag{3.40}$$

Furthermore, $K_{\Sigma}^{\hat{\mathbf{x}}_0}$ *is said to be a stable kernel representation (SKR) of* $\Sigma^{\mathbf{x}_0}$ *if for* $\forall \hat{\mathbf{x}}_0 \in \hat{\mathcal{X}}_0$,

$$\mathbf{y} \in \mathcal{Y}^s, \quad \mathbf{u} \in \mathcal{U}^s \implies K_{\Sigma}^{\hat{\mathbf{x}}_0} \begin{pmatrix} \mathbf{u} \\ \mathbf{y} \end{pmatrix} \in \mathcal{R}^s. \tag{3.41}$$

Definition 3.9. *An image representation of an operator* $\Sigma^{\mathbf{x}_0} : \mathcal{U} \to \mathcal{Y}$ *is any operator* $I_{\Sigma}^{\mathbf{x}_0} : \mathcal{V} \to \mathcal{U} \times \mathcal{Y}$, *such that* $\forall \mathbf{v}, \mathbf{x}_0$

$$I_{\Sigma}^{\mathbf{x}_0}(\mathbf{v}) = \begin{pmatrix} \mathbf{u} \\ \mathbf{y} \end{pmatrix} = \begin{pmatrix} \mathbf{u} \\ \Sigma^{\mathbf{x}_0}(\mathbf{u}) \end{pmatrix}. \tag{3.42}$$

Furthermore, $I_{\Sigma}^{\mathbf{x}_0}$ *is said to be a stable image representation (SIR) of* $\Sigma^{\mathbf{x}_0}$ *if for* $\forall \mathbf{x}_0$

$$\mathbf{v} \in \mathcal{V}^s \implies \mathbf{u} \in \mathcal{U}^s, \quad \mathbf{y} \in \mathcal{Y}^s. \tag{3.43}$$

It is worth mentioning that by means of an SKR, it is able to define an operator for an output observer as follows: $\hat{Y}_{\Sigma}^{\hat{\mathbf{x}}_0} : \mathcal{U}^s \times \mathcal{Y}^s \to \hat{\mathcal{Y}}^s$

$$\hat{Y}_{\Sigma}^{\hat{\mathbf{x}}_0} : \hat{\mathbf{y}} = \mathbf{y} - K_{\Sigma}^{\hat{\mathbf{x}}_0} \begin{pmatrix} \mathbf{u} \\ \mathbf{y} \end{pmatrix}. \tag{3.44}$$

The SKR and SIR of $\Sigma^{\mathbf{x}_0}$ are two alternative description forms of $\Sigma^{\mathbf{x}_0}$ and both of them are stable operators. It follows directly from the SKR and SIR definitions that

$$K_{\Sigma}^{\mathbf{x}_0} \circ I_{\Sigma}^{\mathbf{x}_0} = \mathbf{0}. \tag{3.45}$$

The following definition is needed for introducing the inverses of $K_{\Sigma}^{\hat{\mathbf{x}}_0}, I_{\Sigma}^{\mathbf{x}_0}$.

Definition 3.10. *The SKR* $K_{\Sigma}^{\hat{\mathbf{x}}_0}$ *is said to be coprime if it has a stable right inverse* $K_{\Sigma}^{-} : \mathcal{Z}^s \to \mathcal{U}^s \times \mathcal{Y}^s$ *satisfying*

$$K_{\Sigma}^{\hat{\mathbf{x}}_0} \circ K_{\Sigma}^{-} = \mathbf{I}. \tag{3.46}$$

Analogous, the SIR $I_\Sigma^{\mathbf{x}_0}$ is said to be coprime if it has a stable left inverse $I_\Sigma^- : \mathcal{U}^s \times \mathcal{Y}^s \rightarrow \mathcal{L}^s$ satisfying

$$I_\Sigma^- \circ I_\Sigma^{\mathbf{x}_0} = \mathbf{I}. \tag{3.47}$$

Remark 3.4. *The definition of the SKR has been introduced in [106]. The definition of the SIR is a dual form of SKR, which is closely related to the definition of RCF of a nonlinear operator (system) [106, 105].*

We are now in a position to present the parametrization of nonlinear residual generators. Let $\Sigma_{Q_f}^{\mathbf{x}_q,0} : \mathcal{Z}^s \rightarrow \mathcal{R}^s$, $\Sigma_{Q_f}^{\mathbf{x}_q,0} \neq \mathbf{0}$, be a stable system operator that satisfies $\Sigma_{Q_f}^{\mathbf{x}_q,0}(0) = 0$. Consider the cascade connection $\Sigma_{Q_f}^{\mathbf{x}_q,0} \circ K_\Sigma^{\hat{\mathbf{x}}_0}$. Since in the fault-free case for $\hat{\mathbf{x}}_0 = \mathbf{x}_0$

$$\mathbf{z} = K_\Sigma^{\hat{\mathbf{x}}_0} \begin{pmatrix} \mathbf{u} \\ \mathbf{y} \end{pmatrix} = \mathbf{0} \tag{3.48}$$

we have

$$\Sigma_{Q_f}^{\mathbf{x}_q,0} \circ K_\Sigma^{\hat{\mathbf{x}}_0} \begin{pmatrix} \mathbf{u} \\ \mathbf{y} \end{pmatrix} = \Sigma_{Q_f}^{\mathbf{x}_q,0}(\mathbf{0}) = \mathbf{0}. \tag{3.49}$$

On the other hand, for a detectable fault $\mathbf{w}, \mathbf{y} = \Sigma_f^{\mathbf{x}_0}(\mathbf{u}, \mathbf{w}) \neq \Sigma^{\mathbf{x}_0}(\mathbf{u})$, it leads to

$$\mathbf{z} = K_\Sigma^{\hat{\mathbf{x}}_0} \begin{pmatrix} \mathbf{u} \\ \mathbf{y} \end{pmatrix} \neq K_\Sigma^{\hat{\mathbf{x}}_0} \begin{pmatrix} \mathbf{u} \\ \Sigma^{\mathbf{x}_0}(\mathbf{u}) \end{pmatrix} = \mathbf{0} \Longrightarrow$$
$$\Sigma_{Q_f}^{\mathbf{x}_q,0} \circ K_\Sigma^{\hat{\mathbf{x}}_0} \begin{pmatrix} \mathbf{u} \\ \mathbf{y} \end{pmatrix} = \Sigma_{Q_f}^{\mathbf{x}_q,0}(\mathbf{z}) \neq \mathbf{0}. \tag{3.50}$$

Thus, according to Definition 3.7, $\Sigma_{Q_f}^{\mathbf{x}_q,0} \circ K_\Sigma^{\hat{\mathbf{x}}_0}$ builds a residual generator. In the following theorem, it is shown that any stable residual generator can be parametrized by the cascade connection $\Sigma_{Q_f}^{\mathbf{x}_q,0} \circ K_\Sigma^{\hat{\mathbf{x}}_0}$.

Theorem 3.4. *Let $K_\Sigma^{\hat{\mathbf{x}}_0}$ be the SKR of $\Sigma^{\mathbf{x}_0}$ and $\Sigma_{Q_f}^{\mathbf{x}_q,0}$ be the post-filter defined above. Then any stable residual generator $R_\Sigma^{\xi_0}$ can be parameterized by*

$$R_\Sigma^{\xi_0} = \Sigma_{Q_f}^{\mathbf{x}_q,0} \circ K_\Sigma^{\hat{\mathbf{x}}_0}. \tag{3.51}$$

Proof. According to the definitions of residual generators, SKR and SIR, it holds, in the fault-free case

$$\forall \mathbf{u}, R_{\Sigma}^{\xi_0} \begin{pmatrix} \mathbf{u} \\ \mathbf{y} \end{pmatrix} = \mathbf{0} \Longrightarrow \forall l, R_{\Sigma}^{\xi_0} \circ I_{\Sigma}^{\mathbf{x}_0}(l) = \mathbf{0} \tag{3.52}$$

$$\Longrightarrow R_{\Sigma}^{\xi_0} \circ I_{\Sigma}^{\mathbf{x}_0} = \mathbf{0}. \tag{3.53}$$

Since $K_{\Sigma}^{\hat{\mathbf{x}}_0}, I_{\Sigma}^{\mathbf{x}_0}$ are the coprime kernel and image representations, we have stable $I_{\Sigma}^{-}, K_{\Sigma}^{-}$, and (3.45), (3.46) as well as (3.47) hold. Now, consider $I_{\Sigma}^{-} \circ K_{\Sigma}^{-} : \mathcal{Z} \to \mathcal{L}$. If $I_{\Sigma}^{-} \circ K_{\Sigma}^{-} \neq 0$, there exists an operator $\Pi : \mathcal{Z} \to \mathcal{L}$ such that

$$K_{\Sigma}^{-}(\mathbf{z}) = I_{\Sigma}^{\mathbf{x}_0} \circ \Pi(\mathbf{z}) \Longrightarrow K_{\Sigma}^{-} = I_{\Sigma}^{\mathbf{x}_0} \circ \Pi. \tag{3.54}$$

On the other hand, it follows from (3.45) that

$$K_{\Sigma}^{\mathbf{x}_0} \circ I_{\Sigma}^{\mathbf{x}_0} \circ \Pi = \mathbf{0} \Longrightarrow K_{\Sigma}^{\mathbf{x}_0} \circ K_{\Sigma}^{-} = \mathbf{0} \tag{3.55}$$

which is evidently a contradiction to (3.46). As a result, it can be concluded that

$$I_{\Sigma}^{-} \circ K_{\Sigma}^{-} = \mathbf{0}. \tag{3.56}$$

Eqs. (3.45), (3.46), (3.47) and (3.56) build the Bezout identity, which ensures that the operator

$$\begin{pmatrix} I_{\Sigma}^{-} \\ K_{\Sigma}^{\hat{\mathbf{x}}_0} \end{pmatrix} : \mathcal{U} \times \mathcal{Y} \to \mathcal{L} \times \mathcal{Z} \tag{3.57}$$

has a stable inverse (operator)

$$\begin{pmatrix} I_{\Sigma}^{\mathbf{x}_0} & K_{\Sigma}^{-} \end{pmatrix} : \mathcal{L} \times \mathcal{Z} \to \mathcal{U} \times \mathcal{Y}. \tag{3.58}$$

Re-write $R_{\Sigma}^{\xi_0} \begin{pmatrix} \mathbf{u} \\ \mathbf{y} \end{pmatrix}$ in (3.52) as

$$R_{\Sigma}^{\xi_0} \begin{pmatrix} \mathbf{u} \\ \mathbf{y} \end{pmatrix} = R_{\Sigma}^{\xi_0} \circ \begin{pmatrix} I_{\Sigma}^{\mathbf{x}_0} & K_{\Sigma}^{-} \end{pmatrix} \begin{pmatrix} \mathbf{l} \\ \mathbf{z} \end{pmatrix}. \tag{3.59}$$

Considering (3.53), it yields

$$R_{\Sigma}^{\xi_0} \begin{pmatrix} \mathbf{u} \\ \mathbf{y} \end{pmatrix} = R_{\Sigma}^{\xi_0} \circ K_{\Sigma}^{-}(\mathbf{z}) = R_{\Sigma}^{\xi_0} \circ K_{\Sigma}^{-} \circ K_{\Sigma}^{\hat{\mathbf{x}}_0} \begin{pmatrix} \mathbf{u} \\ \mathbf{y} \end{pmatrix}. \tag{3.60}$$

Setting

$$\Sigma_{Q_f}^{\mathbf{x}_{q,0}} = R_\Sigma^{\xi_0} \circ K_\Sigma^-$$ (3.61)

gives the final result

$$R_\Sigma^{\xi_0} = \Sigma_{Q_f}^{\mathbf{x}_{q,0}} \circ K_\Sigma^{\hat{\mathbf{x}}_0}.$$ (3.62)

Thus the theorem is proved. □

Note that in the cascade configuration $\Sigma_{Q_f}^{\mathbf{x}_{q,0}} \circ K_\Sigma^{\hat{\mathbf{x}}_0}$, $K_\Sigma^{\hat{\mathbf{x}}_0}$ is determined by the system Σ^{x_0} under consideration. In against, the post-filter $\Sigma_{Q_f}^{\mathbf{x}_{q,0}}$ is a stable system and can be arbitrarily constructed. In this sense, $\Sigma_{Q_f}^{\mathbf{x}_{q,0}}$ is understood as a parameter operator (system) and the cascade configuration is called parametrization form of nonlinear residual generators.

Remark 3.5. *In [1], the parametrization form (3.51) has been applied for the purpose of FD system optimization without proving that any residual generator can be parameterized by (3.51).*

As mentioned previously, residual generator $K_\Sigma^{\hat{\mathbf{x}}_0}$ is a kernel representation of Σ^{x_0}. Thus, for a given stable system $\Sigma_{Q_f}^{\mathbf{x}_{q,0}}$, the output of $\Sigma_{Q_f}^{\mathbf{x}_{q,0}} \circ K_\Sigma^{\hat{\mathbf{x}}_0}$ is a residual vector satisfying (3.39). In this sense, it can be concluded that all residual generators can be written into (parameterized as) the general form given by (3.51), which is identical with the parameterization form of residual generators for LTI systems. It is worth to noting that the general residual generator form consists of two sub-systems: a post-filter $\Sigma_{Q_f}^{\mathbf{x}_{q,0}}$ and a kernel presentation $K_\Sigma^{\hat{\mathbf{x}}_0}$. In this context, $K_\Sigma^{\hat{\mathbf{x}}_0}$ is constructed to provide the preliminary form of residual signal while $\Sigma_{Q_f}^{\mathbf{x}_{q,0}}$ is adjusted to obtain significant characteristics of faults. As a result, the configuration of observer-based fault detection system is described in Fig. 3.1, which is composed of kernel representation $K_\Sigma^{\hat{\mathbf{x}}_0}$, post-filter $\Sigma_{Q_f}^{\mathbf{x}_{q,0}}$, evaluation function, threshold setting and decision logic.

3.4 Parametrization of Nonlinear FD Systems

In this section, the parametrization of observer-based FD systems is addressed in terms of the state-space representation. For this purpose, the parametrization form of the residual generators is first described in the state-space configuration, and then the threshold settings corresponding to the two types of residual evaluation functions, $J_E(\mathbf{r})$ and $J_2(\mathbf{r})$, will

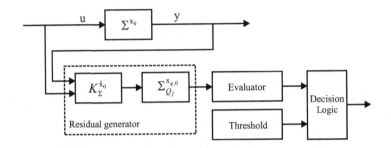

Figure 3.1: Observer-based fault detection configuration for nonlinear systems

be characterized. In this context, nonlinear observer-based FD systems will be parameterized.

Consider the parametrization form of nonlinear residual generators given in Theorem 3.4. Suppose that the state-space representation of the SKR $K_\Sigma^{\hat{x}_0}$ is of the following form

$$K_\Sigma^{\hat{x}_0} : \begin{cases} \dot{\hat{x}} = \phi(\hat{x}, u, y) \\ z = \varphi(\hat{x}, u, y), \end{cases} \quad \hat{x}(0) = \hat{x}_0 \tag{3.63}$$

where $\hat{x} \in \mathcal{R}^n, z \in \mathcal{R}^m$. Recall that

$$\hat{y} = y - \varphi(\hat{x}, u, y) \tag{3.64}$$

delivers an estimate for y. Since for any initial condition $\hat{x}(0) = x(0)$, we have $\varphi(\hat{x}, u, y) = 0 \implies \hat{y} = h(\hat{x}, u)$. Thus, it is reasonable to rewrite z as

$$z = y - \hat{y} = y - h(\hat{x}, u). \tag{3.65}$$

Note that

$$\dot{\hat{x}} = \phi(\hat{x}, u, y), \quad \hat{y} = h(\hat{x}, u) \tag{3.66}$$

is an output observer. System (3.63) with $\varphi(\hat{x}, u, y) = y - \hat{y} = y - h(\hat{x}, u)$ can be called nonlinear FDF.

Let the state-space form of $\Sigma_{Q_f}^{x_{q,0}}$ be

$$\Sigma_{Q_f}^{x_{q,0}} : \begin{cases} \dot{x}_q = f_q(x_q, z) \\ r = h_q(x_q), \end{cases} \quad x_q(0) = x_{q,0}. \tag{3.67}$$

Then, the state space representation of all observer-based residual generators is given by

$$\Sigma_R^{(\hat{\mathbf{x}}_0,\mathbf{x}_{q,0})} : \begin{cases} \dot{\hat{\mathbf{x}}} = \phi(\hat{\mathbf{x}}, \mathbf{u}, \mathbf{y}), \hat{\mathbf{x}}(0) = \hat{\mathbf{x}}_0 \\ \mathbf{z} = \mathbf{y} - \hat{\mathbf{y}} = \mathbf{y} - \mathbf{h}(\hat{\mathbf{x}}, \mathbf{u}) \\ \dot{\mathbf{x}}_q = \mathbf{f}_q(\mathbf{x}_q, \mathbf{z}), \mathbf{x}_q(0) = \mathbf{x}_{q,0} \\ \mathbf{r} = \mathbf{h}_q(\mathbf{x}_q). \end{cases} \tag{3.68}$$

It is worth to notice that the kernel relation (3.40) holds only for $\hat{\mathbf{x}}_0 = \mathbf{x}_0$. In real applications, the state vector $\mathbf{x}$ is not available. Hence, $\hat{\mathbf{x}}_0$ is in general different from $\mathbf{x}_0$. Consequently, the output of the kernel system is different from zero and depends on $\mathbf{x}(0) - \hat{\mathbf{x}}(0) = \mathbf{x}_0 - \hat{\mathbf{x}}_0$ as well as on the process input vector $\mathbf{u}$. For the FD purpose, a residual evaluation function and, associated with it, a threshold are needed to avoid false alarms [30]. This is in fact the motivation for our subsequent study on the state space realization of the parametrization configuration of the residual generators presented in Theorem 3.4 and on the estimation of the possible influence of $\mathbf{x}(0) - \hat{\mathbf{x}}(0)$ and $\mathbf{u}$ on the residual vector. The latter leads to the threshold setting.

In the next two subsections, the parametrizations of the threshold settings for the $\mathcal{L}_\infty$-class and $\mathcal{L}_2$-class FD systems will be addressed, respectively.

3.4.1 $\mathcal{L}_\infty$-Class FD Systems

To this end, two definitions which serve the characterization of the $\mathcal{L}_\infty$-class FD systems are recalled first.

Definition 3.11. *A nonlinear system*

$$\Sigma_{Q_f}^{\mathbf{x}_{q,0}} : \begin{cases} \dot{\mathbf{x}}_q = \mathbf{f}_q(\mathbf{x}_q, \mathbf{z}) \\ \mathbf{r} = \mathbf{h}_q(\mathbf{x}_q), \end{cases} \quad \mathbf{x}_q(0) = \mathbf{x}_{q,0} \tag{3.69}$$

is said to be IOS if there exist functions $\beta(\cdot, t) \in \mathcal{KL}$ and $\sigma(\cdot) \in \mathcal{K}$ such that

$$||\mathbf{r}(t)|| \le \beta(||\mathbf{x}_{q,0}||, t) + \sigma(||\mathbf{z}||_\infty), \, t \ge 0. \tag{3.70}$$

The following results follow immediately from Definitions 3.11 and 3.3.

Theorem 3.5. *Assume that system (3.1) is $\mathcal{L}_\infty$ re-constructible and the post-filter (3.67) is IOS. Let $\mathbf{x}_q(0) = \mathbf{x}_{q,0} = \mathbf{0}$. Then, there exists $\gamma(\cdot) \in \mathcal{K}$ so that $\forall t \geq 0$*

$$||\mathbf{r}(t)|| \leq \gamma(||\mathbf{x}_0 - \hat{\mathbf{x}}_0||) \implies J_{\text{th,E}} = (\gamma(\delta_o))^2 \qquad (3.71)$$

where $\delta_o = \max ||\mathbf{x}_0 - \hat{\mathbf{x}}_0||$.

Proof. It follows from the IOS definition that for $\mathbf{x}_q(0) = \mathbf{x}_{q,0} = \mathbf{0}, t \geq 0$

$$||\mathbf{r}(t)|| \leq \sigma(||\mathbf{z}||_\infty). \qquad (3.72)$$

Since the output re-constructability of the system (3.1) ensures the existence of a $\mathcal{KL}$-function $\beta(||\mathbf{x}_0 - \hat{\mathbf{x}}_0||, t)$ so that

$$
\begin{aligned}
||\mathbf{z}(t)|| &\leq \beta(||\mathbf{x}_0 - \hat{\mathbf{x}}_0||, t) \leq \max_{||\mathbf{x}(0) - \hat{\mathbf{x}}(0)|| \leq \delta_o} \beta(||\mathbf{x}_0 - \hat{\mathbf{x}}_0||, t) \\
&= \beta(\delta_o, 0) = ||z||_\infty. \qquad (3.73)
\end{aligned}
$$

Let $\gamma(\delta_o) = \sigma(\beta(\delta_o, 0))$. It turns out that $\forall t \geq 0$

$$J_{\text{E}}(\mathbf{r}) = ||\mathbf{r}(t)||^2 \leq (\gamma(\delta_o))^2 := J_{\text{th,E}}. \qquad (3.74)$$

The theorem is thus proven. $\qquad\qquad\qquad\qquad\qquad\qquad\qquad$ $\square$

Theorem 3.5 reveals that under certain conditions the threshold can be parameterized by the parameter function γ, a $\mathcal{K}$-function, and the boundedness of the initial state estimation error δ_o, as shown in (3.71).

Applying the existing results on the existence conditions of IOS and $\mathcal{L}_\infty$ re-constructibility of Theorem 3.5 leads to the following corollary.

Corollary 3.2. *Given the system (3.1), the post-filter (3.67) and suppose that there exist (i) a function $\phi : \mathcal{R}^n \times \mathcal{R}^p \times \mathcal{R}^m \to \mathcal{R}^n$ (ii) functions $V(\mathbf{x}, \hat{\mathbf{x}}), \varphi_i(\cdot) \in \mathcal{K}, i = 1, 2, 3$, and constants $\delta, \delta_{\mathbf{u}} > 0$ such that $\forall \mathbf{x}, \hat{\mathbf{x}} \in \mathcal{B}_\delta, ||\mathbf{u}||_\infty \leq \delta_{\mathbf{u}}$*

$$\varphi_1(||\mathbf{y} - \hat{\mathbf{y}}||) \leq V(\mathbf{x}, \hat{\mathbf{x}}) \leq \varphi_2(||\mathbf{x} - \hat{\mathbf{x}}||)$$
$$V_{\mathbf{x}}(\mathbf{x}, \hat{\mathbf{x}})\mathbf{f}(\mathbf{x}, \mathbf{u}) + V_{\hat{\mathbf{x}}}(\mathbf{x}, \hat{\mathbf{x}})\phi(\hat{\mathbf{x}}, \mathbf{u}, \mathbf{y}) \leq -\varphi_3(||\mathbf{x} - \hat{\mathbf{x}}||) \qquad (3.75)$$

(iii) $V_q(\mathbf{x}_q) : \mathcal{R}^{n_q} \to \mathcal{R}_+, \alpha_1(\cdot), \alpha_2(\cdot) \in \mathcal{K}_\infty$ *as well as* $\chi(\cdot) \in \mathcal{K}, \alpha_3(\cdot) \in \mathcal{KL}$ *such that*

$$\alpha_1(\|\mathbf{h}(\mathbf{x}_q)\|) \leq V_q(\mathbf{x}_q) \leq \alpha_2(\|\mathbf{x}_q\|), \forall \mathbf{x}_q \in \mathcal{R}^{n_q}$$
$$V_q(\mathbf{x}_q) \geq \chi(\|\mathbf{z}\|) \implies$$
$$\frac{\partial V_q(\mathbf{x}_q)}{\partial \mathbf{x}_q} \mathbf{f}_q(\mathbf{x}_q, \mathbf{z}) \leq -\alpha_3\left(V_q(\mathbf{x}_q), \|\mathbf{x}_q\|\right). \tag{3.76}$$

Then, there exists $\gamma(\cdot) \in \mathcal{K}$ *so that* $\forall t \geq 0$

$$\|\mathbf{r}(t)\| \leq \gamma\left(\|\mathbf{x}_0 - \hat{\mathbf{x}}_0\|\right). \tag{3.77}$$

Proof. It is proven in [124] that condition (iii) is necessary and sufficient for the stable post-filter (3.67) to be IOS. In [151], it has been demonstrated that if conditions (i) and (ii) hold, then the system under consideration is $\mathcal{L}_\infty$ re-constructible. In fact, the proof is straightforward and can be achieved along with the line in [91, 107] and by applying the result given in [123]. As a result, we have (3.77), as proven in Theorem 3.5. $\qquad\square$

3.4.2 $\mathcal{L}_2$-Class FD System

In the sequel, $\mathcal{L}_2$-class FD systems will be studied. To this end, the well-known definition of $\mathcal{L}_2$-stable systems [134] is reviewed first.

Definition 3.12. *A nonlinear system*

$$\Sigma_{Q_f}^{\mathbf{x}_{q,0}} : \begin{cases} \dot{\mathbf{x}}_q = \mathbf{f}_q(\mathbf{x}_q, \mathbf{z}) \\ \mathbf{r} = \mathbf{h}_q(\mathbf{x}_q), \end{cases} \quad \mathbf{x}_q(0) = \mathbf{x}_{q,0} \tag{3.78}$$

is said to be $\mathcal{L}_2$*-stable if for some constant* $\gamma \geq 0$

$$\|\mathbf{r}\|_{2,\tau}^2 \leq \gamma^2 \|\mathbf{z}\|_{2,\tau}^2 + \gamma_o(\mathbf{x}_{q,0}) \tag{3.79}$$

where $\gamma_o \geq 0$ *is a (finite) constant for a given* $\mathbf{x}_{q,0}$.

For the characterization of $\mathcal{L}_2$-class FD systems with the parametrization configuration given in Theorem 3.4 we have the following result.

Theorem 3.6. *Given the observer-based residual generator (3.68) and assume that the post-filter (3.67) is $\mathcal{L}_2$-stable with $\mathbf{x}_{q,0} = \mathbf{0}$. If there exist functions $V(\mathbf{x}, \hat{\mathbf{x}}) \geq 0, \varphi_1(\cdot) \in \mathcal{K}_\infty$ and a constant $\delta > 0$ such that $\forall \mathbf{x}, \hat{\mathbf{x}} \in \mathcal{B}_\delta$*

$$V_\mathbf{x}(\mathbf{x}, \hat{\mathbf{x}})\mathbf{f}(\mathbf{x}, \mathbf{u}) + V_{\hat{\mathbf{x}}}(\mathbf{x}, \hat{\mathbf{x}})\boldsymbol{\phi}(\hat{\mathbf{x}}, \mathbf{u}, \mathbf{y}) \leq - \|\mathbf{z}\|^2 + \varphi_1(\|\mathbf{u}\|)$$
$$\mathbf{z} = \mathbf{y} - \mathbf{h}(\hat{\mathbf{x}}, \mathbf{u}) \tag{3.80}$$

then it holds

$$
\begin{aligned}
J_2(\mathbf{r}) &= \|\mathbf{r}\|_{2,\tau}^2 \leq \gamma^2 \int_0^\tau \varphi_1(\|\mathbf{u}(t)\|)dt + \gamma_0 \Longrightarrow \\
J_{\mathrm{th},2} &= \gamma^2 \int_0^\tau \varphi_1(\|\mathbf{u}(t)\|)dt + \gamma_0 \\
\gamma_0 &= \gamma^2 \max_{\mathbf{x}_0, \hat{\mathbf{x}}_0 \in \mathcal{B}_\delta} V(\mathbf{x}_0, \hat{\mathbf{x}}_0). \tag{3.81}
\end{aligned}
$$

Proof. It follows from (4.40) that

$$\int_0^\tau \|\mathbf{z}\|^2 \, dt \leq \int_0^\tau \varphi_1(\|\mathbf{u}\|)dt + V(\mathbf{x}(0), \hat{\mathbf{x}}(0)). \tag{3.82}$$

Since (3.67) is $\mathcal{L}_2$-stable with $\mathbf{x}_{q,0} = \mathbf{0}$, it turns out that

$$J_2(\mathbf{r}) = \|\mathbf{r}\|_{2,\tau}^2 \leq \gamma^2 \int_0^\tau \|\mathbf{z}\|^2 \, dt$$
$$\leq \gamma^2 \left(\int_0^\tau \varphi_1(\|\mathbf{u}(t)\|)dt + V(\mathbf{x}(0), \hat{\mathbf{x}}(0)) \right)$$
$$\leq \gamma^2 \int_0^\tau \varphi_1(\|\mathbf{u}(t)\|)dt + \gamma^2 \max_{\mathbf{x}_0, \hat{\mathbf{x}}_0 \in \mathcal{B}_\delta} V(\mathbf{x}_0, \hat{\mathbf{x}}_0). \tag{3.83}$$

As a result, the threshold can be set as

$$J_{\mathrm{th},2} = \gamma^2 \int_0^\tau \varphi_1(\|\mathbf{u}(t)\|)dt + \gamma_0 \tag{3.84}$$

which completes the proof. $\square$

In the FD research, the threshold (3.81) is called adaptive threshold, since it is a function of $\|\mathbf{u}(t)\|$. It is evident that $J_{\mathrm{th},2}$ is parametrized by γ, the $\mathcal{L}_2$-gain of the post-filter, $\varphi_1(\|\mathbf{u}(t)\|)$ and γ_0. It should be pointed out that in the existing studies, for instance in [30, 1, 72], an adaptive threshold is generally parametrized by the $\mathcal{L}_2$-gain of the residual generator. In this work, additional degree of design freedom is introduced in terms of the parameter function φ_1, which can be, for instance, used for the purpose of improving FD performance.

3.5 Numerical Examples

In this section, we illustrate the main results obtained in the previous sections by means of two numerical examples.

Example 3.1: This example is taken from [71] and used to illustrate the results given in Theorem 3.1. Consider nonlinear system (3.1) with

$$\mathbf{x} = \begin{bmatrix} x_1 \\ x_2 \end{bmatrix}, \mathbf{f}(\mathbf{x}, u) = \begin{bmatrix} -x_1^3 + x_2 \\ -x_2^3 + u \end{bmatrix}, h(\mathbf{x}, u) = x_1. \tag{3.85}$$

Suppose that $u \in \mathcal{U} = [-1, 1]$. For $W(\mathbf{x}) = \frac{1}{2}x_1^2 + \frac{1}{2}x_2^2$, it holds $W_\mathbf{x}(\mathbf{x})\mathbf{f}(\mathbf{x}, u) \leq \frac{5}{4} - W^2(\mathbf{x})$. As can be seen from [71], for every initial condition $\mathbf{x}(0) \in \mathcal{R}^2$ and $u \in \mathcal{U} = [-1, 1]$, the solution $\mathbf{x}(t)$ of (3.85) enters the compact set $S = \{\mathbf{x} \in \mathcal{R}^2 : W(\mathbf{x}) \leq \sqrt{10}/2\}$. Now, design the observer-based residual generator as

$$\dot{\hat{x}}_1 = -\hat{x}_1^3 + \hat{x}_2 + l_1(y - \hat{x}_1), \dot{\hat{x}}_2 = -\hat{x}_2^3 + u + l_2(y - \hat{x}_1), r = \hat{x}_1 - y.$$

Let

$$V(\mathbf{x}, \hat{\mathbf{x}}) = \frac{1}{2}(\mathbf{x} - \hat{\mathbf{x}})^T \begin{bmatrix} 1 & -a \\ -a & b \end{bmatrix} (\mathbf{x} - \hat{\mathbf{x}})$$

and denote $e_1 = x_1 - \hat{x}_1, e_2 = x_2 - \hat{x}_2$, it yields

$$V_\mathbf{x}(\mathbf{x}, \hat{\mathbf{x}})\mathbf{f}(\mathbf{x}, u) + V_{\hat{\mathbf{x}}}(\mathbf{x}, \hat{\mathbf{x}})\phi(\hat{\mathbf{x}}, u)$$

$$= \begin{bmatrix} e_1 - ae_2 \\ -ae_1 + be_2 \end{bmatrix}^T \begin{bmatrix} -e_1(\hat{x}_1^2 + x_1^2 + \hat{x}_1 x_1) + e_2 - l_1 e_1 \\ -e_2(\hat{x}_2^2 + x_2^2 + \hat{x}_2 x_2) - l_2 e_1 \end{bmatrix}$$

$$= -e_1^2(\hat{x}_1^2 + x_1^2 + \hat{x}_1 x_1) + ae_2 e_1(\hat{x}_1^2 + x_1^2 + \hat{x}_1 x_1) + e_1 e_2 - l_1 e_1^2$$

$$\quad - ae_2^2 + al_1 e_1 e_2 + ae_1 e_2(\hat{x}_2^2 + x_2^2 + \hat{x}_2 x_2) + al_2 e_1^2 - bl_2 e_1 e_2$$

$$-be_2^2(\hat{x}_2^2 + x_2^2 + \hat{x}_2 x_2)$$

$$\leq ae_1(\hat{x}_1^2 + x_1^2 + \hat{x}_1 x_1 + \hat{x}_2^2 + x_2^2 + \hat{x}_2 x_2)e_2 - ae_2^2 - (l_1 - al_2)e_1^2$$
$$+ (1 + al_1 - bl_2)e_1 e_2$$

$$\leq \frac{a}{2}(\hat{x}_1^2 + x_1^2 + \hat{x}_1 x_1 + \hat{x}_2^2 + x_2^2 + \hat{x}_2 x_2)^2 e_1^2 - \frac{a}{2}e_2^2 - (l_1 - al_2)e_1^2$$
$$+ (1 + al_1 - bl_2)e_1 e_2$$

$$\leq \frac{9a}{8}(x_1^2 + x_2^2 + \hat{x}_1^2 + \hat{x}_2^2)^2 e_1^2 - \frac{a}{2}e_2^2 - (l_1 - al_2)e_1^2 + (1 + al_1 - bl_2)e_1 e_2.$$

Assume that $x_1^2 + x_2^2 \leq c$ and $\hat{x}_1^2 + \hat{x}_2^2 \leq d$, it is easy to verify

$$V_{\mathbf{x}}(\mathbf{x}, \hat{\mathbf{x}})\mathbf{f}(\mathbf{x}, u) + V_{\hat{\mathbf{x}}}(\mathbf{x}, \hat{\mathbf{x}})\phi(\hat{\mathbf{x}}, u)$$

$$\leq (\frac{9a}{8}(c + d)^2 - (l_1 - al_2))e_1^2 - \frac{a}{2}e_2^2 + (1 + al_1 - bl_2)e_1 e_2.$$

Select l_1 and l_2 such that

$$\frac{9a}{8}(c + d)^2 - l_1 + al_2 + \frac{a}{2} = 0, 1 + al_1 - bl_2 = 0$$

holds for arbitrary a, b, d satisfying $b > a^2$, $a > 0$ and $d > c$, then one has that

$$V_{\mathbf{x}}(\mathbf{x}, \hat{\mathbf{x}})\mathbf{f}(\mathbf{x}, u) + V_{\hat{\mathbf{x}}}(\mathbf{x}, \hat{\mathbf{x}})\phi(\hat{\mathbf{x}}, u) \leq -\frac{a}{2}e_1^2 - \frac{a}{2}e_2^2.$$

System (3.85) is output re-constructible for

$$\frac{\lambda_1}{2}r^T r \leq V(\mathbf{x}, \hat{\mathbf{x}}) \leq \frac{\lambda_2}{2}(\mathbf{x} - \hat{\mathbf{x}})^T(\mathbf{x} - \hat{\mathbf{x}})$$

$$V_{\mathbf{x}}(\mathbf{x}, \hat{\mathbf{x}})\mathbf{f}(\mathbf{x}, u) + V_{\hat{\mathbf{x}}}(\mathbf{x}, \hat{\mathbf{x}})\phi(\hat{\mathbf{x}}, u) \leq -\frac{a}{2}(\mathbf{x} - \hat{\mathbf{x}})^T(\mathbf{x} - \hat{\mathbf{x}})$$

where λ_1, λ_2 ($\lambda_2 \geq \lambda_1$) are eigenvalues of $\begin{bmatrix} 1 & -a \\ -a & b \end{bmatrix}$. According to Theorem 3.1, one obtains

$$\dot{V}(\mathbf{x}, \hat{\mathbf{x}}) \leq -\frac{a}{\lambda_2}V(\mathbf{x}, \hat{\mathbf{x}}).$$

It follows from Lemma 6.1 in [123] that

$$\eta(s) = -\int_1^s \frac{\lambda_2 dr}{ar} = -\frac{\lambda_2}{a}\ln(s), \tilde{\gamma}(s, t) = \eta^{-1}(t + \eta(s)) = se^{-\frac{a}{\lambda_2}t}$$

$$\gamma(s, t) = \tilde{\gamma}(s, t) + \frac{s}{t + 1} = se^{-\frac{a}{\lambda_2}t} + \frac{s}{t + 1}.$$

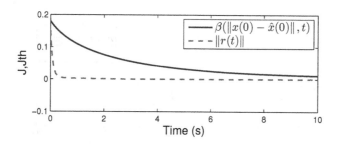

Figure 3.2: The comparison of $\|r(t)\|$ and $\beta(\|\mathbf{x}(0) - \hat{\mathbf{x}}(0)\|, t)$.

Furthermore, it holds that

$$\beta(\|\mathbf{x}(0) - \hat{\mathbf{x}}(0)\|, t) = \frac{2}{\lambda_1}(e^{-\frac{a}{\lambda_2}t} + \frac{1}{t+1})V(\mathbf{x}_0, \hat{\mathbf{x}}_0).$$

In the simulation, $a = 0.5$, $b = 1$, $c = 3.2$, $d = 3.5$ and input $u = 1$ are chosen, which results in $l_1 = 29.4$ and $l_2 = 7.9$. The initial conditions are chosen as $\mathbf{x}(0) = (0.15, 0.15)$ and $\hat{\mathbf{x}}(0) = (0, 0)$. The comparison of $\|r(t)\|$ and $\beta(\|\mathbf{x}(0) - \hat{\mathbf{x}}(0)\|, t)$ is depicted in Fig. 3.2.

Example 3.2: The following example is used to illustrate the effectiveness of Theorem 3.6. Consider a system described by

$$\dot{x}_1 = -x_2 - 2x_1^3 + \frac{1}{2}x_1 u$$

$$\dot{x}_2 = x_1 - x_2^3 - 2x_1^2 x_2$$

$$\dot{x}_3 = -x_1^2 x_3 - 2x_2^2 x_3 - x_3^3$$

$$y = x_1^2 + x_2^2 + x_3^2.$$

Design the observer-based residual generator as

$$\dot{\hat{x}}_1 = -\hat{x}_2 - 2\hat{x}_1^3 + \frac{1}{2}\hat{x}_1 u + \frac{1}{2}\hat{x}_1(y - \hat{y})$$

$$\dot{\hat{x}}_2 = \hat{x}_1 - \hat{x}_2^3 - 2\hat{x}_1^2 \hat{x}_2 + \frac{1}{2}\hat{x}_2(y - \hat{y})$$

$$\dot{\hat{x}}_3 = -2\hat{x}_1^2 \hat{x}_3 - 2\hat{x}_2^2 \hat{x}_3 - \hat{x}_3^3 + \frac{1}{2}\hat{x}_3(y - \hat{y})$$

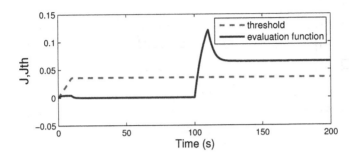

Figure 3.3: Detection of a sensor fault.

$$\hat{y} = \hat{x}_1^2 + \hat{x}_2^2 + \hat{x}_3^2$$
$$r = y - \hat{y}.$$

Let

$$V_{x,\hat{x}} = x_1^2 + x_2^2 + x_3^2 + \hat{x}_1^2 + \hat{x}_2^2 + \hat{x}_3^2$$

it is easy to verify that

$$V_{\mathbf{x}}(\mathbf{x}, \hat{\mathbf{x}})\mathbf{f}(\mathbf{x}, u) + V_{\hat{\mathbf{x}}}(\mathbf{x}, \hat{\mathbf{x}})\phi(\hat{\mathbf{x}}, u, y) \leq -\left\| r \right\|^2 + 0.36 \left\| u \right\|^2.$$

That is to say, the nonlinear system is $\mathcal{L}_2$ re-constructible. It follows from Theorem 3.6 that a $\mathcal{L}_2$ observer-based fault detection system can be realized by setting $J(r) = \|r_\tau\|_2^2$ and $J_{\text{th}} = 0.36 \|u_\tau\|_2^2 + \gamma_o, \gamma_o = \max\limits_{\mathbf{x}(0),\hat{\mathbf{x}}(0)} V(\mathbf{x}(0), \hat{\mathbf{x}}(0))$.

In order to verify the $\mathcal{L}_2$ stability of the NFDF, the input function is chosen as $u = 0.1$. For FD purpose, a constant sensor fault 0.14 is simulated at 100s. With residual evaluation and threshold computation methods provided in Theorem 3.6, it is evident that the fault can be detected as shown in Fig. 3.3. Moreover, by adopting an evaluation window $\tau = 10$s, a dynamic threshold can be attained to improve the fault detection performance.

3.6 Concluding Remarks

In this chapter, observer-based FD issues for nonlinear systems have been addressed. For the analysis purpose of an observer-based FD system, the concepts of $\mathcal{L}_\infty$ type of re-constructability, $\mathcal{L}_2$ type of re-constructability as well as $\mathcal{L}_\infty/\mathcal{L}_2$ type of re-constructability have been introduced. Then it has been proven that (i) if a nonlinear system is $\mathcal{L}_\infty$ re-constructible, an $\mathcal{L}_\infty$ observer-based FD system exists; (ii) if a nonlinear system is $\mathcal{L}_2$ re-constructible, an $\mathcal{L}_2$ observer-based FD system exists; and (iii) if a nonlinear system is $\mathcal{L}_\infty/\mathcal{L}_2$ re-constructible, an $\mathcal{L}_\infty/\mathcal{L}_2$ observer-based FD system exists. Furthermore, the parametrization of nonlinear observer-based FD systems has been addressed. Motivated by the known parametrization scheme for LTI residual generators and the important role of a parametrization in FD system analysis and optimization, the parametrization issues have been studied in two steps. With the aid of nonlinear factorization and input-output operator techniques, it has been first proven that any stable residual generator can be parameterized by a cascade connection of the process kernel representation and a post-filter that represents the parameter system. In the second step, based on the state space representation of the parameterized residual generator, $\mathcal{L}_\infty$- and $\mathcal{L}_2$-class of observer-based FD systems have been investigated. As a result, the threshold settings for both classes of FD systems have been parameterized and, associated with them, the existence conditions have been characterized.

4 Design of $\mathcal{L}_2$ Nonlinear Observer-based FD Systems

The main objective of this chapter is to study the integrated design of $\mathcal{L}_2$ observer-based FD systems for general nonlinear systems. It has been demonstrated in Chapter 3 that $\mathcal{L}_2$ re-constructability serves as the existence condition for $\mathcal{L}_2$ nonlinear observer-based FD systems. For application of these results, a mathematical and systematic tool is needed to handle nonlinear issues. As mentioned in Chapter 1, T-S fuzzy-model-based analysis and synthesis technique is a simple and effective tool to deal with nonlinear systems [126, 156, 18, 83]. Motivated by these observations, in this chapter, the integrated design scheme of $\mathcal{L}_2$ nonlinear observer-based FD systems is addressed with the aid of fuzzy dynamic modelling technique and fuzzy Lyapunov functions. To be specific, by including the input variables into the antecedent part of the rules, the general nonlinear plant is firstly expressed by a set of generalized T-S fuzzy models with norm-bounded approximation errors. Then a fuzzy observer-based residual generator is developed and based on it, a dynamic/adaptive threshold is proposed to give an efficient, real-time, FD system. Moreover, the robust fuzzy observer-based FD systems are investigated for nonlinear processes with disturbances.

4.1 Preliminaries and Problem Formulation

In this chapter, the following type of nonlinear systems is considered

$$\Sigma^{\mathbf{x}_0} : \begin{cases} \dot{\mathbf{x}} = \mathbf{f}(\mathbf{x}, \mathbf{u}) \\ \mathbf{y} = \mathbf{h}(\mathbf{x}, \mathbf{u}) \end{cases} \tag{4.1}$$

where $\mathbf{x} \in \mathcal{R}^{k_x}, \mathbf{y} \in \mathcal{R}^{k_y}, \mathbf{u} \in \mathcal{R}^{k_u}$ denote the state, output and input vectors, respectively. $\mathbf{f}(\mathbf{x}, \mathbf{u})$ and $\mathbf{h}(\mathbf{x}, \mathbf{u})$ are continuously differentiable nonlinear functions with appropriate dimensions.

It is noteworthy to emphasize that Theorem 3.2 provides a condition for checking the existence of an $\mathcal{L}_2$ observer-based FD system for a general type of nonlinear processes (4.1). It presents an analytical framework, which can not be directly applied for design. For the implementation, further efforts are needed, for instance, to find the functions $\varphi_1(\cdot), \varphi_2(\cdot), \phi(\hat{\mathbf{x}}, \mathbf{u}, \mathbf{y})$ and $V(\mathbf{x}, \hat{\mathbf{x}})$. This motivates us to seek the solution for the integrated design scheme of $\mathcal{L}_2$ observer-based FD systems for nonlinear processes (4.1) by solving the condition given in Theorem 3.2. Inspired by the T-S fuzzy dynamic modelling technique proposed in [155] and its application to controller design [47] for general type of nonlinear systems, the next objective is to address the integrated design of nonlinear $\mathcal{L}_2$ observer-based FD systems by applying fuzzy technique as a solution tool. To this end, the T-S fuzzy dynamic model is constructed and its corresponding approximation capacity for general nonlinear systems (4.1) is investigated. Then, the $\mathcal{L}_2$ fuzzy observer-based residual generator will be proposed via fuzzy Lyapunov functions. As a result, an integrated FD system will be constructed. Meanwhile, it is worthy to mention that the solvability of the weakly output re-constructible condition will be addressed by virtue of LMIs. Furthermore, a numerical example will be used to show the advantages of fuzzy Lyapunov-function-based approach over common Lyapunov-function-based approach.

4.2 Design of $\mathcal{L}_2$ Fuzzy Observer-based FD Systems

In this section, the T-S fuzzy dynamic modelling technique is applied to study the $\mathcal{L}_2$ observer-based FD issues for the general type of nonlinear systems (4.1).

4.2.1 Fuzzy Dynamic Modelling

The following class of generalized T-S fuzzy models are employed to approximate nonlinear systems (4.1) first [155, 47]:

Plant Rule $\mathfrak{R}^i$: IF $\theta_1(t)$ is N_1^i and $\theta_2(t)$ is N_2^i and $\cdots$ and $\theta_p(t)$ is N_p^i

$$THEN \begin{cases} \dot{\mathbf{x}}(t) = \mathbf{A}_i\mathbf{x}(t) + \mathbf{B}_i\mathbf{u}(t) \\ \mathbf{y}(t) = \mathbf{C}_i\mathbf{x}(t) + \mathbf{D}_i\mathbf{u}(t), \quad i \in \{1, 2, \cdots, \kappa\} \end{cases} \qquad (4.2)$$

where $\Re^i$ represents the ith fuzzy inference rule; κ denotes the number of inference rules; $\theta(t) = [\theta_1(t) \cdots \theta_p(t)]$ denotes the premise variables assumed to be measurable; $N_j^i (j = 1, 2, \cdots, p)$ indicates the fuzzy sets; $\mathbf{A}_i$, $\mathbf{B}_i$, $\mathbf{C}_i$ and $\mathbf{D}_i$ are system matrices with appropriate dimensions; $\mathbf{x}(t)$, $\mathbf{u}(t)$ and $\mathbf{y}(t)$ denote the system state, input and output variables, respectively.

Let $\mu_i(\theta(t))$ be the normalized fuzzy membership function, which is defined as

$$\mu_i(\theta(t)) = \frac{\prod_{j=1}^p \nu_{ij}(\theta_j(t))}{\sum_{j=1}^\kappa \prod_{j=1}^p \nu_{ij}(\theta_j(t))} \tag{4.3}$$

where $\nu_{ij}(\theta_j(t)) \geq 0$ is the grade of membership of $\theta_j(t)$ in N_j^i. Consequently, one has that

$$\mu_i(\theta(t)) \geq 0, \ i = 1, 2, \cdots, \kappa, \ \sum_{i=1}^\kappa \mu_i(\theta(t)) = 1. \tag{4.4}$$

For ease of presentation, μ_i is denoted as $\mu_i(\theta(t))$ in the sequel.

Thus, by using a center average defuzzifier, a singleton fuzzifier and product inference, the T-S fuzzy system in (4.2) can be inferred as follows:

$$\begin{cases} \dot{\mathbf{x}}(t) = \bar{\mathbf{f}}(\mathbf{x}, \mathbf{u}) \\ \mathbf{y}(t) = \bar{\mathbf{h}}(\mathbf{x}, \mathbf{u}) \end{cases} \tag{4.5}$$

where

$$\bar{\mathbf{f}}(\mathbf{x}, \mathbf{u}) = \sum_{i=1}^\kappa \mu_i \left(\mathbf{A}_i \mathbf{x}(t) + \mathbf{B}_i \mathbf{u}(t) \right)$$

$$\bar{\mathbf{h}}(\mathbf{x}, \mathbf{u}) = \sum_{i=1}^\kappa \mu_i \left(\mathbf{C}_i \mathbf{x}(t) + \mathbf{D}_i \mathbf{u}(t) \right). \tag{4.6}$$

Remark 4.1. *It is noted that if the control variables are excluded from the premise variables, the commonly used T-S fuzzy models can only be adopted to approximate the so-called affine nonlinear systems [155]. Therefore, in order to obtain the T-S fuzzy models for more general types of nonlinear systems (4.1), the control variables $\mathbf{u}(t)$ should be included in the premise variables $\theta(t)$.*

In what follows, the approximation capability of the T-S fuzzy models in (4.2) will be addressed based on the results given in [155, 47].

Theorem 4.1. *Consider nonlinear systems given in (4.1), where* $\mathbf{f}(\mathbf{x}, \mathbf{u})$ *and* $\mathbf{g}(\mathbf{x}, \mathbf{u})$ *are continuously differentiable on the compact set* $\mathcal{X} \times \mathcal{U}$ *and* $\mathbf{f}(0,0) = 0, \mathbf{g}(0,0) = 0$. *Then, for any positive constant* ϵ_i, $i = 1, \cdots, 4$ *and any* $(\mathbf{x}, \mathbf{u}) \in \mathcal{X} \times \mathcal{U}$, *there exist T-S fuzzy models (4.5) such that*

$$\mathbf{f}(\mathbf{x}, \mathbf{u}) = \bar{\mathbf{f}}(\mathbf{x}, \mathbf{u}) + \Delta_{\mathbf{A}}(\mathbf{x}, \mathbf{u})\mathbf{x} + \Delta_{\mathbf{B}}(\mathbf{x}, \mathbf{u})\mathbf{u}$$
$$\mathbf{h}(\mathbf{x}, \mathbf{u}) = \bar{\mathbf{h}}(\mathbf{x}, \mathbf{u}) + \Delta_{\mathbf{C}}(\mathbf{x}, \mathbf{u})\mathbf{x} + \Delta_{\mathbf{D}}(\mathbf{x}, \mathbf{u})\mathbf{u} \qquad (4.7)$$

with

$$\|\Delta_{\mathbf{A}}(\mathbf{x}, \mathbf{u})\| \leq \epsilon_1, \quad \|\Delta_{\mathbf{B}}(\mathbf{x}, \mathbf{u})\| \leq \epsilon_2$$
$$\|\Delta_{\mathbf{C}}(\mathbf{x}, \mathbf{u})\| \leq \epsilon_3, \quad \|\Delta_{\mathbf{D}}(\mathbf{x}, \mathbf{u})\| \leq \epsilon_4. \qquad (4.8)$$

The proof is given in [47].

Remark 4.2. *In industrial applications, model uncertainties always exist, since information about the process is generally insufficient. In this context, model uncertainties can be lumped into the approximation errors (4.8).*

Since fuzzy systems have been shown to be universal approximators for nonlinear systems by a smooth "blending" of a group of local linear models through fuzzy membership functions, we will investigate the fuzzy observer-based residual generator and the integrated FD system for nonlinear processes (4.1) in the foregoing sections by taking into account of the approximation errors of the T-S fuzzy models (4.5).

4.2.2 $\mathcal{L}_2$ Fuzzy Observer-based Residual Generator

The following full-order T-S fuzzy fault detection filter is adopted as a residual generator for nonlinear systems (4.1):

Fuzzy Observer-based Residual Generator Rule $\mathfrak{R}^i$: *IF* $\theta_1(t)$ *is* N_1^i *and* $\theta_2(t)$ *is* N_2^i *and* $\cdots$ *and* $\theta_p(t)$ *is* N_p^i

$$THEN \begin{cases} \dot{\hat{\mathbf{x}}}(t) = \mathbf{A}_i\hat{\mathbf{x}}(t) + \mathbf{B}_i\mathbf{u}(t) + \mathbf{L}_i\left(\mathbf{y}(t) - \hat{\mathbf{y}}(t)\right) \\ \hat{\mathbf{y}}(t) = \mathbf{C}_i\hat{\mathbf{x}}(t) + \mathbf{D}_i\mathbf{u}(t) \\ \mathbf{r}(t) = \mathbf{y}(t) - \hat{\mathbf{y}}(t), \quad i \in \{1, 2, \cdots, \kappa\} \end{cases} \qquad (4.9)$$

where $\hat{\mathbf{x}}(t) \in \mathcal{R}^{k_x}$ denotes the state estimation. $\mathbf{L}_i, i \in \{1, 2, \cdots, \kappa\}$ represents the observer gain matrix to be designed for each local model.

$\mathbf{r}(t) \in \mathcal{R}^{k_y}$ is the so-called residual signal which carries the most important information for fault detection. Then, the overall T-S fuzzy residual generator can be inferred similarly in the following form

$$\dot{\hat{\mathbf{x}}}(t) = \sum_{i=1}^{\kappa} \mu_i \left(\mathbf{A}_i \hat{\mathbf{x}}(t) + \mathbf{B}_i \mathbf{u}(t) + \mathbf{L}_i (\mathbf{y}(t) - \hat{\mathbf{y}}(t)) \right)$$

$$\hat{\mathbf{y}}(t) = \sum_{i=1}^{\kappa} \mu_i \left(\mathbf{C}_i \hat{\mathbf{x}}(t) + \mathbf{D}_i \mathbf{u}(t) \right)$$

$$\mathbf{r}(t) = \mathbf{y}(t) - \hat{\mathbf{y}}(t). \tag{4.10}$$

Defining the estimation error $\mathbf{e}(t) = \mathbf{x}(t) - \hat{\mathbf{x}}(t)$ and setting $\bar{\mathbf{x}}(t) = \left[\mathbf{e}^T(t) \ \mathbf{x}^T(t) \right]^T$, one obtains

$$\dot{\bar{\mathbf{x}}}(t) = \sum_{i=1}^{\kappa} \sum_{j=1}^{\kappa} \mu_i \mu_j \left((\mathcal{A}_{ij} + \Delta_{\mathcal{A}_i}) \bar{\mathbf{x}}(t) + (\mathcal{B}_i + \Delta_{\mathcal{B}_i}) \mathbf{u}(t) \right)$$

$$\mathbf{r}(t) = \sum_{i=1}^{\kappa} \mu_i \left((\mathcal{C}_i + \Delta_{\mathcal{C}}) \bar{\mathbf{x}}(t) + \Delta_{\mathcal{D}} \mathbf{u}(t) \right) \tag{4.11}$$

where

$$\mathcal{A}_{ij} = \begin{bmatrix} \mathbf{A}_i - \mathbf{L}_i \mathbf{C}_j & \mathbf{0} \\ \mathbf{0} & \mathbf{A}_i \end{bmatrix}, \ \mathcal{B}_i = \begin{bmatrix} \mathbf{0} \\ \mathbf{B}_i \end{bmatrix}, \ \mathcal{C}_i = \begin{bmatrix} \mathbf{C}_i & \mathbf{0} \end{bmatrix}$$

$$\Delta_{\mathcal{A}_i} = \begin{bmatrix} \mathbf{0} & \Delta_A(\mathbf{x}, \mathbf{u}) - \mathbf{L}_i \Delta_C(\mathbf{x}, \mathbf{u}) \\ \mathbf{0} & \Delta_A(\mathbf{x}, \mathbf{u}) \end{bmatrix}, \ \Delta_{\mathcal{D}} = \Delta_D(\mathbf{x}, \mathbf{u})$$

$$\Delta_{\mathcal{B}_i} = \begin{bmatrix} \Delta_B(\mathbf{x}, \mathbf{u}) - \mathbf{L}_i \Delta_D(\mathbf{x}, \mathbf{u}) \\ \Delta_B(\mathbf{x}, \mathbf{u}) \end{bmatrix}, \ \Delta_{\mathcal{C}} = \begin{bmatrix} \mathbf{0} & \Delta_C(\mathbf{x}, \mathbf{u}) \end{bmatrix}.$$

The following theorem provides a design scheme for the determination of gain matrices $\mathbf{L}_i$, $i = 1, \cdots, \kappa$, via fuzzy Lyapunov functions, which also serves as the tool to check the weakly output re-constructible condition for systems (4.1).

Theorem 4.2. *Assume that*

$$|\dot{\mu}_\rho(\theta(t))| \leq \eta_\rho, \quad \rho = 1, \cdots, \kappa - 1. \tag{4.12}$$

Given nonlinear system (4.1) and fuzzy residual generator (4.10). Suppose that there exist constants $\alpha > 0, \xi > 0$ and matrices $\mathbf{Z}_1, \mathbf{Z}_2, \mathbf{L}_i, \mathbf{P}_i >$

0, $i = 1, \cdots, \kappa$, such that the following inequalities hold

$$\mathbf{P}_\rho \geq \mathbf{P}_\kappa, \quad \rho = 1, \cdots, \kappa - 1 \tag{4.13}$$

$$\Xi_{ii} < \mathbf{0}, \quad i = 1, \cdots, \kappa \tag{4.14}$$

$$\Xi_{ij} + \Xi_{ji} < \mathbf{0}, \quad 1 \leq i < j \leq \kappa \tag{4.15}$$

where

$$\Xi_{ij} = \begin{bmatrix} \mathbf{\Psi}_{ij} + \mathbf{G}_2 & \star \\ \mathbf{H}_j & -\xi\mathbf{I} \end{bmatrix}, \mathbf{M}_j = \begin{bmatrix} \mathbf{I} & -\mathbf{L}_j \\ \mathbf{I} & \mathbf{0} \end{bmatrix}, \mathbf{N} = \begin{bmatrix} \mathbf{0} & \mathbf{I} \end{bmatrix}$$

$$\mathbf{\Psi}_{ij} = \begin{bmatrix} \mathbb{S} - \mathbf{Z}_1\mathcal{A}_{ij} - \mathcal{A}_{ij}^T\mathbf{Z}_1^T & \star & \star & \star \\ -\mathcal{B}_i^T\mathbf{Z}_1^T & -\alpha^2\mathbf{I} & \star & \star \\ \mathbf{P}_i + \mathbf{Z}_1^T - \mathbf{Z}_2\mathcal{A}_{ij} & -\mathbf{Z}_2\mathcal{B}_i & \mathbf{Z}_2 + \mathbf{Z}_2^T & \star \\ \mathcal{C}_i & \mathbf{0} & \mathbf{0} & -\mathbf{I} \end{bmatrix}$$

$$\mathbf{G}_2 = \begin{bmatrix} \begin{bmatrix} \mathbf{0} & \star \\ \mathbf{0} & \xi\lambda\mathbf{I}_{k_x} \end{bmatrix} & \star & \star & \star \\ \mathbf{0} & \xi\lambda\mathbf{I}_{k_u} & \star & \star \\ \mathbf{0} & \mathbf{0} & \mathbf{0} & \star \\ \mathbf{0} & \mathbf{0} & \mathbf{0} & \mathbf{0} \end{bmatrix}$$

$$\mathbf{H}_j = \begin{bmatrix} -\mathbf{M}_j^T\mathbf{Z}_1^T & \mathbf{0} & -\mathbf{M}_j^T\mathbf{Z}_2^T & \mathbf{N}^T \end{bmatrix}$$

$$\mathbb{S} = \sum_{\rho=1}^{\kappa-1} \eta_\rho(\mathbf{P}_\rho - \mathbf{P}_\kappa), \lambda = \epsilon_1^2 + \epsilon_2^2 + \epsilon_3^2 + \epsilon_4^2. \tag{4.16}$$

Then, it holds that

$$\|\mathbf{r}\|_2^2 < \alpha \|\mathbf{u}\|_2^2 + \sum_{i=1}^{\kappa} \mu_i(\theta(0))\bar{\mathbf{x}}^T(0)\mathbf{P}_i\bar{\mathbf{x}}(0). \tag{4.17}$$

Proof. Consider the following fuzzy Lyapunov candidate function

$$V(\bar{\mathbf{x}}(t)) = \sum_{i=1}^{\kappa} \mu_i\bar{\mathbf{x}}^T(t)\mathbf{P}_i\bar{\mathbf{x}}(t) \tag{4.18}$$

where $\mathbf{P}_i, i = 1, \cdots, \kappa$ are positive definite symmetric matrices. Note that

$$\dot{\mathbf{V}}(\bar{\mathbf{x}}(t)) + \mathbf{r}^T(t)\mathbf{r}(t) - \alpha^2\mathbf{u}^T(t)\mathbf{u}(t) < 0 \tag{4.19}$$

yields (4.17). Thus, in what follows, we are devoted to seek the solvability of (4.19). To this end, we first take the time derivative of $\mathbf{V}(\bar{\mathbf{x}}(t))$ along the trajectory of system (4.11)

$$\dot{\mathbf{V}}(\bar{\mathbf{x}}(t)) = 2\sum_{i=1}^{\kappa} \mu_i \dot{\bar{\mathbf{x}}}^T(t)\mathbf{P}_i\bar{\mathbf{x}}(t) + \sum_{i=1}^{\kappa} \dot{\mu}_i \bar{\mathbf{x}}^T(t)\mathbf{P}_i\bar{\mathbf{x}}(t). \tag{4.20}$$

It follows from

$$\sum_{\rho=1}^{\kappa} \dot{\mu}_\rho(\theta(t)) = 0, \ \dot{\mu}_\kappa(\theta(t)) = -\sum_{\rho=1}^{\kappa-1} \dot{\mu}_\rho(\theta(t)), \ \forall \theta(t) \tag{4.21}$$

and (4.12)-(4.13) that

$$\sum_{\rho=1}^{\kappa} \dot{\mu}_\rho \mathbf{P}_\rho = \sum_{\rho=1}^{\kappa-1} \dot{\mu}_\rho(\mathbf{P}_\rho - \mathbf{P}_\kappa) \le \sum_{\rho=1}^{\kappa-1} \eta_\rho(\mathbf{P}_\rho - \mathbf{P}_\kappa) = \mathbb{S}. \tag{4.22}$$

Moreover, it is evident that for any non-singular matrices $\mathbf{Z}_1$ and $\mathbf{Z}_2$ with appropriate dimensions, one has that

$$2\sum_{i=1}^{\kappa}\sum_{j=1}^{\kappa} \mu_i\mu_j(\bar{\mathbf{x}}^T(t)\mathbf{Z}_1 + \dot{\bar{\mathbf{x}}}^T(t)\mathbf{Z}_2)$$
$$\times \left(\dot{\bar{\mathbf{x}}}(t) - (\mathcal{A}_{ij} + \Delta_{\mathcal{A}_i})\bar{\mathbf{x}}(t) - (\mathcal{B}_i + \Delta_{\mathcal{B}_i})\mathbf{u}(t)\right) = 0. \tag{4.23}$$

Hence, by defining $\mathbf{z}^T(t) = \begin{bmatrix} \bar{\mathbf{x}}^T(t) & \mathbf{u}^T(t) & \dot{\bar{\mathbf{x}}}^T(t) \end{bmatrix}$ and considering (4.22)-(4.23), it becomes evident that the following inequality implies (4.19)

$$\sum_{i=1}^{\kappa}\sum_{j=1}^{\kappa} \mu_i\mu_j \mathbf{z}^T(t)\mathbf{\Theta}_{ij}\mathbf{z}(t) + \mathbf{r}^T(t)\mathbf{r}(t) < 0 \tag{4.24}$$

where

$$\mathbf{\Theta}_{ij} = \begin{bmatrix} \mathbb{S} - \mathrm{Sym}\{\mathbf{Z}_1\mathbb{A}_{ij}\} & \star & \star \\ -\mathbb{B}_i^T\mathbf{Z}_1^T & -\alpha^2\mathbf{I} & \star \\ \mathbf{P}_i + \mathbf{Z}_1^T - \mathbf{Z}_2\mathbb{A}_{ij} & -\mathbf{Z}_2\mathbb{B}_i & \mathbf{Z}_2 + \mathbf{Z}_2^T \end{bmatrix}$$
$$\mathbb{A}_{ij} = \mathcal{A}_{ij} + \Delta_{\mathcal{A}_i}, \ \mathbb{B}_i = \mathcal{B}_i + \Delta_{\mathcal{B}_i}. \tag{4.25}$$

By Schur Complement, it is easy to see that the following inequality leads to (4.24)

$$\sum_{i=1}^{\kappa}\sum_{j=1}^{\kappa} \mu_i \mu_j \mathbf{\Pi}_{ij} < 0 \qquad (4.26)$$

where

$$\mathbf{\Pi}_{ij} = \mathbf{\Psi}_{ij} + \mathbf{H}_j \Delta_{\mathbf{\Psi}} + \Delta_{\mathbf{\Psi}}^T \mathbf{H}_j^T$$
$$\Delta_{\mathbf{\Psi}} = \begin{bmatrix} \begin{bmatrix} 0 & \Delta_\mathbf{A}(\mathbf{x}, \mathbf{u}) \\ 0 & \Delta_\mathbf{C}(\mathbf{x}, \mathbf{u}) \end{bmatrix} & \Delta_\mathbf{B}(\mathbf{x}, \mathbf{u}) & 0 & 0 \\ & \Delta_\mathbf{D}(\mathbf{x}, \mathbf{u}) & 0 & 0 \end{bmatrix}. \qquad (4.27)$$

Here, $\mathbf{H}_j, j = 1, \cdots, \kappa$ and $\mathbf{\Psi}_{ij}, i, j = 1, \cdots, \kappa$ are defined in (4.16). Noting that (4.26) can be equivalently expressed as

$$\sum_{i=1}^{\kappa}\sum_{j>i}^{\kappa} \mu_i \mu_j \left(\mathbf{\Pi}_{ij} + \mathbf{\Pi}_{ji} \right) + \sum_{i=1}^{\kappa} \mu_i^2 \mathbf{\Pi}_{ii} < 0 \qquad (4.28)$$

thus it is easy to see that (4.26) is fulfilled providing that the following inequalities are feasible

$$\mathbf{\Pi}_{ii} < 0, \quad i = 1, \cdots, \kappa \qquad (4.29)$$
$$\mathbf{\Pi}_{ij} + \mathbf{\Pi}_{ji} < 0, \quad i = 1, \cdots, \kappa, \ j > i. \qquad (4.30)$$

For ease of presentation, we only focus on the proof of the more complicated case (4.30). Note that for any positive constant ξ, we have

$$(\mathbf{H}_i + \mathbf{H}_j)\Delta_{\mathbf{\Psi}} + \Delta_{\mathbf{\Psi}}^T (\mathbf{H}_i + \mathbf{H}_j)^T$$
$$\leq \frac{1}{2\xi}(\mathbf{H}_i + \mathbf{H}_j)(\mathbf{H}_i + \mathbf{H}_j)^T + 2\xi \Delta_{\mathbf{\Psi}}^T \Delta_{\mathbf{\Psi}}. \qquad (4.31)$$

In addition, it can be easily proved that for $\lambda = \epsilon_1^2 + \epsilon_2^2 + \epsilon_3^2 + \epsilon_4^2$, one has that

$$\begin{bmatrix} \Delta_\mathbf{A}(\mathbf{x}, \mathbf{u}) & \Delta_\mathbf{B}(\mathbf{x}, \mathbf{u}) \\ \Delta_\mathbf{C}(\mathbf{x}, \mathbf{u}) & \Delta_\mathbf{D}(\mathbf{x}, \mathbf{u}) \end{bmatrix}^T (\star) \leq \lambda \mathbf{I}. \qquad (4.32)$$

Furthermore, by Schur Complement, it is easy to see that (4.30) holds if (4.15) is feasible. Thus, the proof is completed. $\qquad \square$

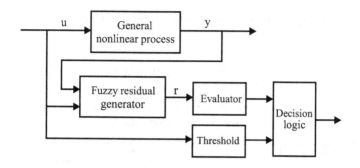

Figure 4.1: Fuzzy observer-based FD configuration for nonlinear systems

Remark 4.3. *It is noted that by setting $\mathbf{P}_i = \mathbf{P} > 0, i = 1, \cdots , \kappa$, the determination of the gain matrices can be realized via the common Lyapunov-function-based approach. The comparison between these two approaches will be addressed by an example in Section 4.4.*

Remark 4.4. *Note that (4.7) is the fuzzy dynamic model of (4.1) and (4.18) and (4.19) are the variants of $V(\mathbf{x}, \hat{\mathbf{x}})$ and (3.22) in Theorem 3.2, respectively. Thus, it can be concluded that the solvability of (4.13)-(4.15) imply (3.22) with $\varphi_1(\|\mathbf{r}\|) = \mathbf{r}^T \mathbf{r}$ and $\varphi_2(\|\mathbf{u}\|) = \alpha^2 \mathbf{u}^T \mathbf{u}$, which means the nonlinear system (4.1) is $\mathcal{L}_2$ re-constructible. Moreover, in order to solve (4.14)-(4.15) with the aid of convex optimization, let*

$$\mathbf{Z}_1 = \left[\begin{array}{cc} \mathbf{Z}_{11} & \mathbf{Z}_{12} \\ 0 & \mathbf{Z}_{13} \end{array} \right], \mathbf{Z}_2 = \left[\begin{array}{cc} \mathbf{Z}_{11} & \mathbf{Z}_{22} \\ 0 & \mathbf{Z}_{23} \end{array} \right] \tag{4.33}$$

and $\mathbf{Q}_j = \mathbf{Z}_{11}\mathbf{L}_j$, then (4.14)-(4.15) can be rewritten into a set of LMIs [80].

4.2.3 $\mathcal{L}_2$ Fuzzy Observer-based FD Systems

Based on the residual generator proposed in Theorem 4.2, an $\mathcal{L}_2$ fuzzy observer-based FD system for nonlinear processes (4.1) can be realized (as shown in Fig. 4.1) by (i) defining

$$J(\mathbf{r}) = \|\mathbf{r}_\tau\|_2^2 := \int_{t_0}^{t_0+\tau} \mathbf{r}^T(t)\mathbf{r}(t)dt \tag{4.34}$$

(ii) determining the associated dynamic/adaptive threshold

$$J_{\text{th}} = \alpha^2 \left\| \mathbf{u}_\tau \right\|_2^2 + \gamma_o$$

$$\gamma_o = \max_{\mathbf{x}(0),\hat{\mathbf{x}}(0)} \sum_{i=1}^{\kappa} \mu_i(\theta(0)) \bar{\mathbf{x}}^T(0) \mathbf{P}_i \bar{\mathbf{x}}(0). \tag{4.35}$$

Then the on-line realization of $\mathcal{L}_2$ fuzzy observer-based FD method for nonlinear systems is summarized in Algorithm 1.

Algorithm 1 On-line FD algorithm for nonlinear processes

1: Run the residual generator (4.10)
2: Run the evaluator (4.34)
3: Set the adaptive threshold (4.35)
4: Run the decision logic

$$\begin{cases} J(\mathbf{r}) = \left\| \mathbf{r}_\tau \right\|_2^2 > J_{\text{th}} \implies \text{faulty} \\ J(\mathbf{r}) = \left\| \mathbf{r}_\tau \right\|_2^2 \leq J_{\text{th}} \implies \text{fault-free} \end{cases} \tag{4.36}$$

4.3 Design of $\mathcal{L}_2$ Robust Fuzzy Observer-based FD Systems

In this section, we address the robust fuzzy observer-based FD issues for nonlinear systems with external disturbances, which is described by

$$\Sigma^{\mathbf{x}_0} : \begin{cases} \dot{\mathbf{x}} = \mathbf{f}(\mathbf{x}, \mathbf{u}) + \mathbf{g}(\mathbf{x}, \mathbf{u})\mathbf{d} \\ \mathbf{y} = \mathbf{h}(\mathbf{x}, \mathbf{u}) + \mathbf{k}(\mathbf{x}, \mathbf{u})\mathbf{d} \end{cases} \tag{4.37}$$

where $\mathbf{g}(\mathbf{x}, \mathbf{u})$ and $\mathbf{k}(\mathbf{x}, \mathbf{u})$ are continuously differentiable nonlinear function matrices with appropriate dimensions. $\mathbf{d} \in \mathcal{R}^{k_d}$ denotes the disturbances which is $\mathcal{L}_2$-bounded with

$$\left\| \mathbf{d}_\tau \right\|_2 \leq \delta_{\mathbf{d}}. \tag{4.38}$$

Analog to the previous section, the nonlinear systems (4.37) can be approximated by the following T-S fuzzy models with some norm bounded uncertainties:

Plant rule $\mathcal{R}^i$: **IF** $\theta_1(t)$ is N_1^i and $\theta_2(t)$ is N_2^i and $\cdots$ and $\theta_p(t)$ is N_p^i

$$
\textbf{THEN} \begin{cases} \dot{\mathbf{x}}(t) = \mathbf{A}_i\mathbf{x}(t) + \mathbf{B}_i\mathbf{u}(t) + \mathbf{E}_i\mathbf{d}(t) + \Delta_{\mathbf{A}}(\mathbf{x},\mathbf{u})\mathbf{x}(t) \\ \quad +\Delta_{\mathbf{B}}(\mathbf{x},\mathbf{u})\mathbf{u}(t) + \Delta_{\mathbf{E}}(\mathbf{x},\mathbf{u})\mathbf{d}(t) \\ \mathbf{y}(t) = \mathbf{C}_i\mathbf{x}(t) + \mathbf{D}_i\mathbf{u}(t) + \mathbf{F}_i\mathbf{d}(t) + \Delta_{\mathbf{C}}(\mathbf{x},\mathbf{u})\mathbf{x}(t) \\ \quad +\Delta_{\mathbf{D}}(\mathbf{x},\mathbf{u})\mathbf{u}(t) + \Delta_{\mathbf{F}}(\mathbf{x},\mathbf{u})\mathbf{d}(t), \ i \in \{1,2,\cdots,\kappa\} \end{cases}
$$
$$(4.39)$$

where $\mathbf{E}_i$ and $\mathbf{F}_i$ are matrices with appropriate dimensions. It follows from Theorem 7.1 that for any positive ϵ_5, ϵ_6 and any $(\mathbf{x},\mathbf{u}) \in \mathcal{X} \times \mathcal{U}$

$$
\|\Delta_{\mathbf{E}}(\mathbf{x},\mathbf{u})\| \leq \epsilon_5, \ \ \|\Delta_{\mathbf{F}}(\mathbf{x},\mathbf{u})\| \leq \epsilon_6. \tag{4.40}
$$

Thus the final state of the fuzzy system can be inferred as follows:

$$
\dot{\mathbf{x}}(t) = \sum_{i=1}^{\kappa} \mu_i \left(\mathbf{A}_i\mathbf{x}(t) + \mathbf{B}_i\mathbf{u}(t) + \mathbf{E}_i\mathbf{d}(t) \right) + \Delta_{\mathbf{A}}(\mathbf{x},\mathbf{u})\mathbf{x}(t)
$$
$$
+ \Delta_{\mathbf{B}}(\mathbf{x},\mathbf{u})\mathbf{u}(t) + \Delta_{\mathbf{E}}(\mathbf{x},\mathbf{u})\mathbf{d}(t)
$$
$$
\mathbf{y}(t) = \sum_{i=1}^{\kappa} \mu_i \left(\mathbf{C}_i\mathbf{x}(t) + \mathbf{D}_i\mathbf{u}(t) + \mathbf{F}_i\mathbf{d}(t) \right) + \Delta_{\mathbf{C}}(\mathbf{x},\mathbf{u})\mathbf{x}(t)
$$
$$
+ \Delta_{\mathbf{D}}(\mathbf{x},\mathbf{u})\mathbf{u}(t) + \Delta_{\mathbf{F}}(\mathbf{x},\mathbf{u})\mathbf{d}(t). \tag{4.41}
$$

By adopting the fuzzy observer-based residual generator of the form (4.9), the overall T-S fuzzy residual generator can be further represented as (4.10).

Denote

$$
\mathbf{e}(t) = \mathbf{x}(t) - \hat{\mathbf{x}}(t), \ \bar{\mathbf{x}}(t) = \begin{bmatrix} \mathbf{e}^T(t) & \mathbf{x}^T(t) \end{bmatrix}^T, \ \mathbf{z}(t) = \begin{bmatrix} \mathbf{u}^T(t) & \mathbf{d}^T(t) \end{bmatrix}^T
$$

$$
\mathcal{A}_{i,j} = \begin{bmatrix} \mathbf{A}_i - \mathbf{L}_i\mathbf{C}_j & 0 \\ 0 & \mathbf{A}_i \end{bmatrix}, \ \Delta_{\mathcal{A}_i} = \begin{bmatrix} 0 & \Delta_{\mathbf{A}}(\mathbf{x},\mathbf{u}) - \mathbf{L}_i\Delta_{\mathbf{C}}(\mathbf{x},\mathbf{u}) \\ 0 & \Delta_{\mathbf{A}}(\mathbf{x},\mathbf{u}) \end{bmatrix}
$$

$$
\mathcal{B}_i = \begin{bmatrix} 0 & \mathbf{E}_i \\ \mathbf{B}_i & \mathbf{E}_i \end{bmatrix}, \ \mathcal{C}_i = \begin{bmatrix} \mathbf{C}_i & 0 \end{bmatrix}, \ \Delta_{\mathcal{C}} = \begin{bmatrix} 0 & \Delta_{\mathbf{C}}(\mathbf{x},\mathbf{u}) \end{bmatrix}
$$

$$
\mathcal{D}_i = \begin{bmatrix} 0 & \mathbf{F}_i \end{bmatrix}, \ \Delta_{\mathcal{D}} = \begin{bmatrix} \Delta_{\mathbf{D}}(\mathbf{x},\mathbf{u}) & \Delta_{\mathbf{F}}(\mathbf{x},\mathbf{u}) \end{bmatrix}
$$

$$
\Delta_{\mathcal{B}_i} = \begin{bmatrix} \Delta_{\mathbf{B}}(\mathbf{x},\mathbf{u}) - \mathbf{L}_i\Delta_{\mathbf{D}}(\mathbf{x},\mathbf{u}) & \Delta_{\mathbf{E}}(\mathbf{x},\mathbf{u}) - \mathbf{L}_i\Delta_{\mathbf{F}}(\mathbf{x},\mathbf{u}) \\ \Delta_{\mathbf{B}}(\mathbf{x},\mathbf{u}) & \Delta_{\mathbf{E}}(\mathbf{x},\mathbf{u}) \end{bmatrix}
$$

we obtain

$$\dot{\bar{\mathbf{x}}}(t) = \sum_{i=1}^{\kappa} \sum_{j=1}^{\kappa} \mu_i \mu_j \left((\mathcal{A}_{i,j} + \Delta_{\mathcal{A}_i}) \bar{\mathbf{x}}(t) + (\mathcal{B}_i + \Delta_{\mathcal{B}_i}) \mathbf{z}(t) \right)$$

$$\mathbf{r}(t) = \sum_{i=1}^{\kappa} \mu_i \left((\mathcal{C}_i + \Delta_{\mathcal{C}}) \bar{\mathbf{x}}(t) + (\mathcal{D}_i + \Delta_{\mathcal{D}}) \mathbf{z}(t) \right) \tag{4.42}$$

With a slight modification, the major result in Theorem 4.2 can be applied to solve the following (robust) FD problem.

Theorem 4.3. *Given nonlinear systems (4.37) and fuzzy residual generator (4.10). Suppose that there exist constants $\alpha > 0, \xi > 0$ and matrices $\mathbf{Z}_1, \mathbf{Z}_2, \mathbf{L}_i, \mathbf{P}_i > 0$, $i = 1, \cdots, \kappa$, such that the following inequalities hold*

$$\mathbf{P}_\rho \geq \mathbf{P}_\kappa, \quad \rho = 1, \cdots, \kappa - 1 \tag{4.43}$$

$$\Xi_{ii} < 0, \quad i = 1, \cdots, \kappa \tag{4.44}$$

$$\Xi_{ij} + \Xi_{ji} < 0, \quad 1 \leq i < j \leq \kappa \tag{4.45}$$

where

$$\Xi_{ij} = \begin{bmatrix} \boldsymbol{\Psi}_{ij} + \mathbf{G}_2 & \star \\ \mathbf{H}_j & -\xi \mathbf{I} \end{bmatrix}, \mathbf{M}_j = \begin{bmatrix} \mathbf{I} & -\mathbf{L}_j \\ \mathbf{I} & 0 \end{bmatrix}, \mathbf{N} = \begin{bmatrix} 0 & \mathbf{I} \end{bmatrix}$$

$$\boldsymbol{\Psi}_{ij} = \begin{bmatrix} \mathbb{S} - \mathbf{Z}_1 \mathcal{A}_{ij} - \mathcal{A}_{ij}^T \mathbf{Z}_1^T & \star & \star & \star \\ -\mathcal{B}_i^T \mathbf{Z}_1^T & -\alpha^2 \mathbf{I} & \star & \star \\ \mathbf{P}_i + \mathbf{Z}_1^T - \mathbf{Z}_2 \mathcal{A}_{ij} & -\mathbf{Z}_2 \mathcal{B}_i & \mathbf{Z}_2 + \mathbf{Z}_2^T & \star \\ \mathcal{C}_i & \mathcal{D}_i & 0 & -\mathbf{I} \end{bmatrix}$$

$$\mathbf{H}_j = \begin{bmatrix} -\mathbf{M}_j^T \mathbf{Z}_1^T & 0 & -\mathbf{M}_j^T \mathbf{Z}_2^T & \mathbf{N}^T \end{bmatrix}$$

$$\mathbf{G}_2 = \begin{bmatrix} \begin{bmatrix} 0 & \star \\ 0 & \xi \lambda \mathbf{I}_{k_x} \end{bmatrix} & \star & \star & \star \\ 0 & \xi \lambda \mathbf{I}_{k_u + k_d} & \star & \star \\ 0 & 0 & 0 & \star \\ 0 & 0 & 0 & 0 \end{bmatrix}$$

$$\mathbb{S} = \sum_{\rho=1}^{\kappa-1} \eta_\rho (\mathbf{P}_\rho - \mathbf{P}_\kappa), \lambda = \epsilon_1^2 + \epsilon_2^2 + \epsilon_3^2 + \epsilon_4^2 + \epsilon_5^2 + \epsilon_6^2. \tag{4.46}$$

Then, it holds that

$$\|\mathbf{r}\|_2^2 < \alpha^2 \|\mathbf{z}\|_2^2 + \sum_{i=1}^{\kappa} \mu_i(\theta(0))\bar{x}^T(0)\mathbf{P}_i\bar{\mathbf{x}}(0). \qquad (4.47)$$

Proof. The proof of this theorem is quite similar to Theorem 4.2 and thus is omitted here. $\qquad\square$

As a result, the following scheme can be applied to the realization of an FD system:

- Run the residual generator (4.10)

- Set

$$J_{\text{th}} = \alpha \left(\|\mathbf{u}_\tau\|_2^2 + \delta_{\mathbf{d}}^2 \right) + \sup_{\mathbf{x}(0),\hat{\mathbf{x}}(0)} \mathbf{V}(\bar{\mathbf{x}}(0)) \qquad (4.48)$$

- Set the detection logic

$$\begin{cases} J(\mathbf{r}) = \|\mathbf{r}_\tau\|_2^2 > J_{\text{th}} \Longrightarrow \text{faulty} \\ J(\mathbf{r}) = \|\mathbf{r}_\tau\|_2^2 \le J_{\text{th}} \Longrightarrow \text{fault-free} \end{cases} \qquad (4.49)$$

4.4 A Numerical Example

In this example, we demonstrate the advantage of fuzzy Lyapunov-function-based approach over common Lyapunov-function-based approach. Consider the following fuzzy system:

Plant rule $\Re^i$: IF $u(t)$ is N_1^i, THEN

$$\begin{cases} \dot{\mathbf{x}} = \mathbf{A}_i\mathbf{x} + \mathbf{B}_i u + \Delta_{\mathbf{A}}(\mathbf{x},u)\mathbf{x} + \Delta_{\mathbf{B}}(\mathbf{x},u)u \\ \mathbf{y} = \mathbf{C}_i\mathbf{x} + \Delta_{\mathbf{C}}(\mathbf{x})\mathbf{x}, \quad i \in \{1,2\} \end{cases}$$

where

$$\mathbf{A}_1 = \begin{bmatrix} 0 & 1 \\ -\varsigma & -1 \end{bmatrix}, \mathbf{B}_1 = \begin{bmatrix} 1 \\ 1 \end{bmatrix}, \mathbf{C}_1 = \begin{bmatrix} 0.015 & 0.26 \end{bmatrix}$$

$$\mathbf{A}_2 = \begin{bmatrix} 0 & 1 \\ -2 & -1 \end{bmatrix}, \mathbf{B}_2 = \begin{bmatrix} 2 \\ 1 \end{bmatrix}, \mathbf{C}_2 = \begin{bmatrix} 0.025 & 0.26 \end{bmatrix}.$$

Table 4.1: Comparison of FD performance for different approaches

ς	6.1	5.1	4.1	3.1	2.1
α_f	19.70	12.55	7.65	4.33	2.04
α_c	Infeasible	1988	43.23	8.64	2.15

ς is a parameter given in a real domain. It is assumed that $||\Delta_{\mathbf{A}}(\mathbf{x}, u)|| \leq 0.2$, $||\Delta_{\mathbf{B}}(\mathbf{x}, u)|| \leq 0.1$, $||\Delta_{\mathbf{C}}(\mathbf{x})|| \leq 0.1$ and $|\dot{\mu}_1(u(t))| \leq \eta_1 = 0.3$.

For ease of presentation, we denote α_f and α_c as α obtained based on fuzzy and common Lyapunov-function-based approaches, respectively. The comparison results between these two approaches for each value of ς are shown in Table 4.1. It is evident that the fuzzy Lyapunov-function-based approach can be applied to deal with a larger class of nonlinear systems, and, meanwhile, significantly improve the FD performance.

4.5 Concluding Remarks

The main focus of this chapter is on the integrated design of $\mathcal{L}_2$ observer-based FD systems for general nonlinear processes. It is noted that the $\mathcal{L}_2$ re-constructible condition, as the existence condition for an $\mathcal{L}_2$ observer-based FD system, has been studied in Chapter 3. Our work in this chapter is an integrated design by solving the proposed existence condition with the aid of T-S fuzzy dynamic modelling technique. It has been proved that the solvability of $\mathcal{L}_2$ re-constructible condition can be conducted into the solution of a set of LMIs via fuzzy Lyapunov functions. To be specific, the basic idea consists in fuzzy dynamic modelling, fuzzy residual generator and further the integrated FD system. Furthermore, an efficient FD algorithm has been attained by invoking an adaptive threshold. A numerical example has been used to demonstrate the application and advantages of fuzzy Lyapunov function based approach also in comparison with the common Lyapunov function based approach.

5 Design of $\mathcal{L}_\infty/\mathcal{L}_2$ Nonlinear Observer-based FD Systems

From engineering viewpoint, it is of great significance to attain a real-time detection of potential faulty components or subsystems, since failure may result in disastrous consequences and hazards for personnel, plant and environment. Therefore, the investigations on real-time fault detection approaches have attracted increasing attention in the engineering domains [137, 8, 99]. Motivated by these arguments, the major objective of this chapter is to address integrated design issues of a real-time $\mathcal{L}_\infty/\mathcal{L}_2$ type of observer-based FD for a general type of nonlinear industrial processes with external disturbances. It is noted that an existence condition for $\mathcal{L}_\infty/\mathcal{L}_2$ type of nonlinear observer-based FD systems has been proposed in Chapter 3. For a successful application of the results to the FD system design, T-S fuzzy dynamic modelling technique is applied. For this purpose, the nonlinear system is first described by a smooth "blending" of a set of T-S fuzzy models with norm-bounded approximation errors. Then an asynchronous fuzzy residual generator is developed via piecewise Lyapunov functions by handling the proposed design condition. Specifically, this asynchronous algorithm is able to deal with the situation that the premise variables of the residual generator are not as the same as the premise variables of the T-S fuzzy models of the nonlinear plant. Based on the residual generator, an observer-based FD system is proposed with an embedded dynamic threshold. In comparison with the standard nonlinear $\mathcal{L}_2$ observer-based FD approaches [72, 151], the proposed scheme may lead to a significant improvement on the real-time ability of the FD performance.

5.1 Preliminaries and Problem Formulation

In this chapter, the following general type of nonlinear systems is considered

$$\Sigma : \begin{cases} \dot{\mathbf{x}} = \mathbf{f}(\mathbf{x}, \mathbf{u}) + \mathbf{g}(\mathbf{x}, \mathbf{u})\mathbf{d} \\ \mathbf{y} = \mathbf{h}(\mathbf{x}) \end{cases} \tag{5.1}$$

where $\mathbf{x}(t) \in \mathcal{R}^{k_x}$ is the system state; $\mathbf{y}(t) \in \mathcal{R}^{k_y}$ is the measurement output; $\mathbf{u}(t) \in \mathcal{R}^{k_u}$ is the control input. $\mathbf{d}(t) \in \mathcal{R}^{k_d}$ represents the exogenous disturbance signal which is assumed to be $\mathcal{L}_2$-bounded with $\|\mathbf{d}\|_2 \leq \sigma_d$. $\mathbf{f}(\mathbf{x}, \mathbf{u})$, $\mathbf{g}(\mathbf{x}, \mathbf{u})$ and $\mathbf{h}(\mathbf{x})$ are continuously differentiable nonlinear functions with appropriate dimensions.

Without loss of generality, the nonlinear system in case of fault can be modelled as follows:

$$\tilde{\Sigma} : \begin{cases} \dot{\mathbf{x}} = \tilde{\mathbf{f}}(\mathbf{x}, \mathbf{u}, \mathbf{w}) + \tilde{\mathbf{g}}(\mathbf{x}, \mathbf{u}, \mathbf{w})\mathbf{d} \\ \mathbf{y} = \tilde{\mathbf{h}}(\mathbf{x}, \mathbf{w}) \end{cases} \tag{5.2}$$

where $\mathbf{w} \in \mathcal{R}^{k_w}$ denotes the fault vector. $\tilde{\mathbf{f}}(\mathbf{x}, \mathbf{u}, \mathbf{w})$ and $\tilde{\mathbf{h}}(\mathbf{x}, \mathbf{w})$ are continuous nonlinear functions with appropriate dimensions. It is noted that in fault-free case, $\tilde{\mathbf{f}}(\mathbf{x}, \mathbf{u}, 0) = \mathbf{f}(\mathbf{x}, \mathbf{u})$, $\tilde{\mathbf{g}}(\mathbf{x}, \mathbf{u}, 0) = \mathbf{g}(\mathbf{x}, \mathbf{u})$, $\tilde{\mathbf{h}}(\mathbf{x}, 0) = \mathbf{h}(\mathbf{x})$.

In the sequel, the $\mathcal{L}_\infty/\mathcal{L}_2$ re-constructible concept is first generalized to nonlinear systems (5.1).

Definition 5.1. *System (5.1) is said to be $\mathcal{L}_\infty/\mathcal{L}_2$ re-constructible if there exists a nonlinear system*

$$\begin{cases} \dot{\hat{\mathbf{x}}} = \phi(\hat{\mathbf{x}}, \mathbf{u}, \mathbf{y}) \\ \hat{\mathbf{y}} = \mathbf{h}(\hat{\mathbf{x}}) \end{cases} \tag{5.3}$$

such that $\forall \mathbf{x}, \hat{\mathbf{x}} \in \mathcal{B}_\delta$

$$\varphi_1(\|\mathbf{y} - \hat{\mathbf{y}}\|) \leq \int_0^\tau \varphi_2(\|\mathbf{u}(t)\|)dt + \int_0^\tau \varphi_3(\|\mathbf{d}(t)\|)dt + \gamma_o(\mathbf{x}(0), \hat{\mathbf{x}}(0)) \tag{5.4}$$

where $\varphi_1(\cdot) \in \mathcal{K}$, $\varphi_2(\cdot) \in \mathcal{K}_\infty$, $\delta > 0$ and $\gamma_o(\cdot) \geq 0$ is a (finite) constant for given $\mathbf{x}(0), \hat{\mathbf{x}}(0)$.

Similar to Theorem 3.3, a sufficient condition for the $\mathcal{L}_\infty/\mathcal{L}_2$ type of re-constructability is introduced in the following theorem, which also serves as the existence condition for an $\mathcal{L}_\infty/\mathcal{L}_2$ type of observer-based FD system and the threshold setting.

Corollary 5.1. *Given system (5.1), if there exist (i) a function $\phi : \mathcal{R}^{k_x} \times \mathcal{R}^{k_u} \times \mathcal{R}^{k_y} \to \mathcal{R}^{k_x}$; (ii) functions $V(\mathbf{x}, \hat{\mathbf{x}}) : \mathcal{R}^{k_x} \times \mathcal{R}^{k_x} \to \mathcal{R}^+, \varphi_1(\cdot) \in \mathcal{K}, \varphi_2(\cdot) \in \mathcal{K}_\infty, \varphi_3(\cdot) \in \mathcal{K}_\infty$ and a positive constant δ such that $\forall \mathbf{x}, \hat{\mathbf{x}} \in \mathcal{B}_\delta$*

$$\varphi_1(\|\mathbf{r}\|) \le V(\mathbf{x}, \hat{\mathbf{x}}), \quad \mathbf{r} = \mathbf{y} - \hat{\mathbf{y}} \tag{5.5}$$

$$V_\mathbf{x}(\mathbf{x}, \hat{\mathbf{x}})\mathbf{f}(\mathbf{x}, \mathbf{u}) + V_{\hat{\mathbf{x}}}(\mathbf{x}, \hat{\mathbf{x}})\phi(\hat{\mathbf{x}}, \mathbf{u}, \mathbf{h}(\mathbf{x}, \mathbf{u})) \le \varphi_2(\|\mathbf{u}\|) + \varphi_3(\|\mathbf{d}\|) \tag{5.6}$$

then system (5.1) is $\mathcal{L}_\infty/\mathcal{L}_2$ re-constructible. Moreover, an $\mathcal{L}_\infty/\mathcal{L}_2$ type of observer-based FD systems can be realized by (i) constructing residual generator (5.3) (ii) defining the residual evaluation function as

$$J(\mathbf{r}) = \varphi_1(\|\mathbf{r}(t)\|) \tag{5.7}$$

and (iii) setting the threshold

$$J_{\text{th}} = \int_0^\tau \varphi_2\left(\|\mathbf{u}(t)\|\right) dt + \int_0^\tau \varphi_3\left(\|\mathbf{d}(t)\|\right) dt + \bar{\gamma}_0$$

$$\bar{\gamma}_0 = \sup_{\mathbf{x}(0), \hat{\mathbf{x}}(0) \in \mathcal{B}_\delta} V\left(\mathbf{x}(0), \hat{\mathbf{x}}(0)\right). \tag{5.8}$$

A similar proof can be found in Theorem 3.3.

Note that (5.5)-(5.6) only presents an analytical framework, but does not lead to the direct design approach of observer-based FD systems for nonlinear processes. For FD purpose, further efforts are needed, for instance, to find the functions $\varphi_1(\cdot), \varphi_2(\cdot), \varphi_3(\cdot), \phi(\hat{\mathbf{x}}, \mathbf{u}, \mathbf{y})$ and $V(\mathbf{x}, \hat{\mathbf{x}})$. This motivates us to seek the solution for the integrated design scheme of $\mathcal{L}_\infty/\mathcal{L}_2$ type of observer-based FD systems for general nonlinear processes (5.1) with the aid of T-S fuzzy dynamic modelling technique.

5.2 Design of $\mathcal{L}_\infty/\mathcal{L}_2$ Type of Fuzzy Observer-based FD Systems

5.2.1 Fuzzy Dynamic Modelling

As shown in [47, 48, 155], by including the input variable into the premise variables, the following class of generalized T-S fuzzy models can be employed to approximate nonlinear systems (5.1):

Plant rule $\Re^k$: IF $\theta_1(t)$ is N_1^k and $\theta_2(t)$ is N_2^k and $\cdots$ and $\theta_p(t)$ is N_p^k

$$
THEN \begin{cases} \dot{x}(t) = \mathbf{A}_k\mathbf{x}(t) + \mathbf{B}_k\mathbf{u}(t) + \mathbf{E}_k\mathbf{d}(t) + \mathbf{a}_k + \Delta_A(\mathbf{x},\mathbf{u})\mathbf{x} \\ \qquad +\Delta_B(\mathbf{x},\mathbf{u})\mathbf{u} + \Delta_E(\mathbf{x},\mathbf{u})\mathbf{d} \\ \mathbf{y}(t) = \mathbf{C}_k\mathbf{x}(t) + \Delta_C(\mathbf{x})\mathbf{x}, \quad k \in \{1,2,\cdots,\kappa\} \end{cases} \tag{5.9}
$$

where $\Re^k$ represents the kth fuzzy inference rule; κ denotes the number of inference rules; $\theta(t) = [\theta_1(t) \; \cdots \; \theta_p(t)]$ denotes the premise variables assumed measurable; $N_j^l(j = 1, 2, \cdots, p)$ indicates the fuzzy sets; $\mathbf{A}_k, \mathbf{B}_k$, $\mathbf{E}_k$ and $\mathbf{C}_k$ are system matrices of the kth local model with appropriate dimensions; $\mathbf{a}_k$ is the offset; $\mathbf{x}(t), \mathbf{u}(t)$ and $\mathbf{y}(t)$ denote the system state, input and output variables, respectively. It is noted for any given positive constants $\epsilon_1, \epsilon_2, \epsilon_3$ and ϵ_4, there exists T-S fuzzy dynamic model (5.9) for nonlinear systems (5.1) such that

$$
||\Delta_{\mathbf{A}}(\mathbf{x},\mathbf{u})|| \le \epsilon_1, \quad ||\Delta_{\mathbf{B}}(\mathbf{x},\mathbf{u})|| \le \epsilon_2
$$
$$
||\Delta_{\mathbf{C}}(\mathbf{x})|| \le \epsilon_3, \quad ||\Delta_{\mathbf{E}}(\mathbf{x},\mathbf{u})|| \le \epsilon_4. \tag{5.10}
$$

Remark 5.1. *It is noted with the additional offset term $\mathbf{a}_k$, the approximation capabilities of the fuzzy models (5.9) can be improved [18].*

Let $\mu_k(\theta(t))$ represent the corresponding normalized fuzzy membership function as (4.3). by using a singleton fuzzifier, a center average defuzzifier and product inference, the global T-S fuzzy system can be inferred as follows:

$$
\begin{cases} \dot{x}(t) = \mathcal{A}(\mu)\mathbf{x} + \mathcal{B}(\mu)\mathbf{u} + \mathcal{E}(\mu)\mathbf{u} + \mathbf{a}(\mu) + \Delta_A(\mathbf{x},\mathbf{u})\mathbf{x} \\ \qquad +\Delta_B(\mathbf{x},\mathbf{u})\mathbf{u} + \Delta_E(\mathbf{x},\mathbf{u})\mathbf{d} \\ \mathbf{y}(t) = \mathcal{C}(\mu)\mathbf{u} + \Delta_C(\mathbf{x})\mathbf{x} \end{cases} \tag{5.11}
$$

where

$$
\mathcal{A}(\mu) = \sum_{i=1}^{\kappa} \mu_i \mathbf{A}_i, \mathcal{B}(\mu) = \sum_{i=1}^{\kappa} \mu_i \mathbf{B}_i, \mathbf{a}(\mu) = \sum_{i=1}^{\kappa} \mu_i \mathbf{a}_i
$$
$$
\mathcal{E}(\mu) = \sum_{i=1}^{\kappa} \mu_i \mathbf{E}_i, \mathcal{C}(\mu) = \sum_{i=1}^{\kappa} \mu_i \mathbf{C}_i. \tag{5.12}
$$

In the sequel, the design problem of $\mathcal{L}_\infty/\mathcal{L}_2$ type of observer-based FD systems for nonlinear processes (5.1) will be addressed based on piecewise residual generators and piecewise quadratic Lyapunov functions. It is noted that the rules of the fuzzy system induce a polyhedral partition of the state space. In this sense, the global fuzzy system (5.11) can be viewed as a smooth "blending" of a group of local models in a set of individual regions. Let $\{\mathcal{S}_i\}_{i \in \ell}$ denote the state-space partition and ℓ the set of region indices. Along the ideas of [66, 81], we partition the premise variable space $\mathcal{S} \in \mathcal{R}^p$ into two kinds of regions: crisp (operating) regions and fuzzy (interpolation) regions. The crisp region is defined as the region where $\mu_l(\theta(t)) = 1$ for some l. The system dynamics of the crisp region is governed by the lth local model. On the other hand, the fuzzy region is defined as the region where $0 < \mu_l(\theta(t)) < 1$, with the system dynamics described by a blending of several local models. Then, nonlinear systems (5.1) can be equivalently re-written in the following piecewise-fuzzy form:

$$\begin{cases} \dot{\mathbf{x}}(t) = (\mathcal{A}_i + \Delta_{\mathbf{A}}(\mathbf{x}, \mathbf{u}))\, \mathbf{x}(t) + (\mathcal{B}_i + \Delta_{\mathbf{B}}(\mathbf{x}, \mathbf{u}))\, \mathbf{u}(t) \\ \qquad + (\mathcal{E}_i + \Delta_{\mathbf{E}}(\mathbf{x}, \mathbf{u}))\, \mathbf{d}(t) + \mathbf{a}_i \\ \mathbf{y}(t) = (\mathcal{C}_i + \Delta_{\mathbf{C}}(\mathbf{x}))\, \mathbf{x}(t), \theta(t) \in \mathcal{S}_i, \ i \in \ell \end{cases} \tag{5.13}$$

where

$$\mathcal{A}_i = \sum_{l \in \Gamma(i)} \mu_l \mathbf{A}_l, \mathcal{B}_i = \sum_{l \in \Gamma(i)} \mu_l \mathbf{B}_l, \mathbf{a}_i = \sum_{l \in \Gamma(i)} \mu_l \mathbf{a}_l$$

$$\mathcal{E}_i = \sum_{l \in \Gamma(i)} \mu_l \mathbf{E}_l, \mathcal{C}_i = \sum_{l \in \Gamma(i)} \mu_l \mathbf{C}_l \tag{5.14}$$

with $0 < \mu_l(\theta(t)) \leq 1$ and $\sum_{l \in \Gamma(i)} \mu_l(\theta(t)) = 1$. For each region $\mathcal{S}_i$,

$$\Gamma(i) := \{l | \mu_l(\theta(t)) > 0, \ \theta(t) \in \mathcal{S}_i, \ i \in \ell\} \tag{5.15}$$

is defined as the indices for the system matrices used in the interpolation within the region $\mathcal{S}_i$. Separate the partitioned regions into two classes $\ell = \ell_0 \cup \ell_1$, where ℓ_0 represents the set of region indices that contain the origin, and where ℓ_1 represents the set of region indices that do not contain the origin. It is noted that $\mathbf{a}_i = 0$ for all $i \in \ell_0$.

5.2.2 $\mathcal{L}_\infty/\mathcal{L}_2$ Type of Fuzzy Observer-based Residual Generator

Given the piecewise fuzzy system (5.13) on each region, the $\mathcal{L}_\infty/\mathcal{L}_2$ type of fuzzy observer-based FD systems will be addressed in the sequel by solving the proposed design condition (5.6). It is noteworthy to mention that the premise variables of T-S fuzzy models can be partially or completely unmeasurable in some complex industrial processes. To deal with this issue, the following piecewise fuzzy FDF is adopted as a residual generator for nonlinear system (5.1):

Region Rule s: IF $\hat{\theta}(t) \in \mathcal{S}_s, s \in \ell$, THEN

Local Observer-based Residual Generator Rule $\Re^n$: IF $\hat{\theta}_1(t)$ is N_1^n and $\hat{\theta}_2(t)$ is N_2^n and $\cdots$ and $\hat{\theta}_p(t)$ is N_p^n, THEN

$$\begin{cases} \dot{\hat{\mathbf{x}}}(t) = \mathbf{A}_n\hat{\mathbf{x}}(t) + \mathbf{B}_n\mathbf{u}(t) + \mathbf{a}_n + \mathbf{L}_{sn}\left(\mathbf{y}(t) - \hat{\mathbf{y}}(t)\right) \\ \hat{\mathbf{y}}(t) = \mathbf{C}_n\hat{\mathbf{x}}(t) \\ \mathbf{r}(t) = \mathbf{y}(t) - \hat{\mathbf{y}}(t), \quad n \in \Gamma(s) \end{cases} \tag{5.16}$$

where $\hat{\theta}(t) = \begin{bmatrix} \hat{\theta}_1(t) & \cdots & \hat{\theta}_p(t) \end{bmatrix}$ are the premise variables of the residual generator, $\hat{\mathbf{x}}(t) \in R^{k_x}$ is the estimated state; $\mathbf{r}(t) \in R^{k_y}$ is the residual signal; $\mathbf{L}_{sn}, n \in \Gamma(s), s \in \ell$ is the gain matrix to be determined of each local model in each local region.

Similarly, the overall piecewise fuzzy residual generator can be inferred in the following form

$$\begin{cases} \dot{\hat{\mathbf{x}}}(t) = \hat{\mathcal{A}}_s\hat{\mathbf{x}}(t) + \hat{\mathcal{B}}_s\mathbf{u}(t) + \hat{\mathbf{a}}_s + \hat{\mathcal{L}}_s\left(\mathbf{y}(t) - \hat{\mathbf{y}}(t)\right) \\ \hat{\mathbf{y}}(t) = \hat{\mathcal{C}}_s\hat{\mathbf{x}}(t) \\ \mathbf{r}(t) = \mathbf{y}(t) - \hat{\mathbf{y}}(t), \quad \hat{\theta}(t) \in \mathcal{S}_s, \ s \in \ell \end{cases} \tag{5.17}$$

where

$$\hat{\mathcal{A}}_s = \sum_{n\in\Gamma(s)} \hat{\mu}_n(\hat{\theta}(t))\mathbf{A}_n, \hat{\mathcal{B}}_s = \sum_{n\in\Gamma(s)} \hat{\mu}_n(\hat{\theta}(t))\mathbf{B}_n$$

$$\hat{\mathbf{a}}_s = \sum_{n\in\Gamma(s)} \hat{\mu}_n(\hat{\theta}(t))\mathbf{a}_n, \hat{\mathcal{L}}_s = \sum_{n\in\Gamma(s)} \hat{\mu}_n(\hat{\theta}(t))\mathbf{L}_{sn}$$

$$\hat{\mathcal{C}}_s = \sum_{n \in \Gamma(s)} \hat{\mu}_n(\hat{\theta}(t)) \mathbf{C}_n$$

$$\hat{\mu}_n(\hat{\theta}(t)) = \frac{\prod_{j=1}^p \nu_{nj}(\hat{\theta}_j(t))}{\sum_{j=1}^\kappa \prod_{j=1}^p \nu_{nj}(\hat{\theta}_j(t))}. \tag{5.18}$$

Remark 5.2. *Generally speaking, the T-S fuzzy models of the complex highly nonlinear systems are derived via off-line modeling [126, 18]. In this case, the piecewise fuzzy system (5.13) is partitioned based on the general state space, instead of the measurable output space. As a result, it is not always in the situation that the plant and the observer are operating in the same region, especially in the initial stage of the system dynamics. This motivates the investigations on the non-synchronized residual generator synthesis in the case that the premise variables of the fuzzy systems are not measurable. In such circumstances, the proposed scheme can be also applied to the observer-based FD design for nonlinear processes within the network-based environment, since the network-induced limitations such as time delays, quantization, packet dropouts always exists. Some works on the controller/filter design issues for fuzzy systems in presence of unmeasurable premise variables have been reported in [101, 22, 78, 112, 6].*

Next, defining the estimation error $\mathbf{e}(t) = \mathbf{x}(t) - \hat{\mathbf{x}}(t)$ and setting $\eta(t) = \begin{bmatrix} \mathbf{e}^T(t) & \mathbf{x}^T(t) \end{bmatrix}^T$, $\xi(t) = \begin{bmatrix} \mathbf{u}^T(t) & \mathbf{d}^T(t) \end{bmatrix}^T$, the overall augmented dynamics of the residual generators can be described by

$$\begin{cases} \dot{\eta}(t) = \left(\tilde{\mathcal{A}}_{is} + \Delta_{\tilde{\mathbf{A}}_s} \right) \eta(t) + (\tilde{\mathcal{B}}_{is} + \Delta_{\tilde{\mathcal{B}}}) \mathbf{1} \xi(t) + \tilde{\mathbf{a}}_{is} \\ \mathbf{r}(t) = \left(\tilde{\mathcal{C}}_{is} + \Delta_{\tilde{\mathcal{C}}} \right) \bar{\mathbf{x}}(t) \end{cases} \tag{5.19}$$

where

$$\tilde{\mathcal{A}}_{is} = \begin{bmatrix} \hat{\mathcal{A}}_s - \hat{\mathcal{L}}_s \hat{\mathcal{C}}_s & \mathcal{A}_i - \hat{\mathcal{L}}_s \mathcal{C}_i - \hat{\mathcal{A}}_s + \hat{\mathcal{L}}_s \hat{\mathcal{C}}_s \\ 0 & \mathcal{A}_i \end{bmatrix}$$

$$\Delta_{\tilde{\mathcal{A}}_s} = \begin{bmatrix} 0 & \Delta_{\mathbf{A}}(\mathbf{x}, \mathbf{u}) - \hat{\mathcal{L}}_s \Delta_{\mathbf{C}}(\mathbf{x}) \\ 0 & \Delta_{\mathbf{A}}(\mathbf{x}, \mathbf{u}) \end{bmatrix}, \tilde{\mathbf{a}}_{is} = \begin{bmatrix} \mathbf{a}_i - \hat{\mathbf{a}}_s \\ \mathbf{a}_i \end{bmatrix}$$

$$\tilde{\mathcal{B}}_{is} = \begin{bmatrix} \mathcal{B}_i - \hat{\mathcal{B}}_s & \mathcal{D}_i \\ \mathcal{B}_i & \mathcal{D}_i \end{bmatrix}, \Delta_{\tilde{\mathcal{B}}} = \begin{bmatrix} \Delta_{\mathbf{B}}(\mathbf{x}, \mathbf{u}) & \Delta_{\mathbf{D}}(\mathbf{x}, \mathbf{u}) \\ \Delta_{\mathbf{B}}(\mathbf{x}, \mathbf{u}) & \Delta_{\mathbf{D}}(\mathbf{x}, \mathbf{u}) \end{bmatrix}$$

$$\tilde{\mathcal{C}}_{is} = \begin{bmatrix} \hat{\mathcal{C}}_s & \mathcal{C}_i - \hat{\mathcal{C}}_s \end{bmatrix}, \Delta_{\tilde{\mathcal{C}}} = \begin{bmatrix} 0 & \Delta_{\mathbf{C}}(\mathbf{x}) \end{bmatrix}.$$

For convenient notations, it is adopted that

$$\bar{\eta}(t) = \begin{bmatrix} \eta(t) \\ 1 \end{bmatrix}, \bar{\mathcal{A}}_{is} = \begin{bmatrix} \tilde{\mathcal{A}}_{is} + \Delta_{\tilde{\mathcal{A}}_s} & \tilde{a}_{is} \\ 0 & 0 \end{bmatrix}$$

$$\bar{\mathcal{B}}_{is} = \begin{bmatrix} \tilde{\mathcal{B}}_{is} + \Delta_{\tilde{\mathcal{B}}} \\ 0 \end{bmatrix}, \bar{\mathcal{C}}_{is} = \begin{bmatrix} \tilde{\mathcal{C}}_{is} + \Delta_{\tilde{\mathcal{C}}} & 0 \end{bmatrix}. \tag{5.20}$$

Now, system (5.19) can be rewritten as

$$\begin{cases} \dot{\bar{\eta}}(t) = \vec{\mathcal{A}}_{is}\bar{\eta}(t) + \vec{\mathcal{B}}_{is}\xi(t) \\ \mathbf{r}(t) = \vec{\mathcal{C}}_{is}\bar{\eta}(t), \quad \theta(t) \in S_i, \hat{\theta}(t) \in S_s \; i, s \in \ell \end{cases} \tag{5.21}$$

where

$$\begin{bmatrix} \vec{\eta}(t) & | & \vec{\mathcal{A}}_{is} & | & \vec{\mathcal{B}}_{is} & | & \vec{\mathcal{C}}_{is} \end{bmatrix}$$

$$= \begin{cases} \begin{bmatrix} \eta(t) & | & \tilde{\mathcal{A}}_{is} + \Delta_{\tilde{\mathcal{A}}_s} & | & \tilde{\mathcal{B}}_{is} + \Delta_{\tilde{\mathcal{B}}} & | & \tilde{\mathcal{C}}_{is} + \Delta_{\tilde{\mathcal{C}}} \end{bmatrix} \\ \qquad\qquad\qquad\qquad\text{if } i, s \in \ell_0 \\ \begin{bmatrix} \bar{\eta}(t) & | & \bar{\mathcal{A}}_{is} & | & \bar{\mathcal{B}}_{is} & | & \bar{\mathcal{C}}_{is} \end{bmatrix}, \text{ else.} \end{cases} \tag{5.22}$$

In order to find the PQLF that is continuous across the region boundaries, the following continuity matrices $\bar{\mathbf{F}}_i = \begin{bmatrix} \mathbf{F}_i & \mathbf{f}_i \end{bmatrix}, i \in \ell$, with $\mathbf{f}_i = 0$ for $i \in \ell_0$ are constructed to characterize the boundary among the regions:

$$\bar{\mathbf{F}}_i \begin{bmatrix} \mathbf{x} \\ 1 \end{bmatrix} = \bar{\mathbf{F}}_j \begin{bmatrix} \mathbf{x} \\ 1 \end{bmatrix}, \quad \mathbf{x} \in \mathcal{S}_i \cap \mathcal{S}_j$$

$$\bar{\mathbf{F}}_s \begin{bmatrix} \hat{\mathbf{x}} \\ 1 \end{bmatrix} = \bar{\mathbf{F}}_k \begin{bmatrix} \hat{\mathbf{x}} \\ 1 \end{bmatrix}, \quad \hat{\mathbf{x}} \in \mathcal{S}_s \cap \mathcal{S}_k, \; i, j, s, k \in \ell. \tag{5.23}$$

Based on (5.23), one obtains

$$\begin{bmatrix} -\mathbf{F}_s & \mathbf{F}_s & \mathbf{f}_s \\ 0 & \mathbf{F}_i & \mathbf{f}_i \end{bmatrix} \begin{bmatrix} \mathbf{e} \\ \mathbf{x} \\ 1 \end{bmatrix} = \begin{bmatrix} -\mathbf{F}_k & \mathbf{F}_k & \mathbf{f}_k \\ 0 & \mathbf{F}_j & \mathbf{f}_j \end{bmatrix} \begin{bmatrix} \mathbf{e} \\ \mathbf{x} \\ 1 \end{bmatrix}$$

$$\mathbf{x} \in \mathcal{S}_i \cap \mathcal{S}_j, \hat{\mathbf{x}} \in \mathbf{S}_s \cap \mathbf{S}_k. \tag{5.24}$$

As a result, the piecewise quadratic Lyapunov matrices can be parameterized by

$$\vec{\mathbf{P}}_{is} = \vec{\mathbf{F}}_{is}^T T \vec{\mathbf{F}}_{is}, \; i, s \in \ell \tag{5.25}$$

where T is symmetric matrix and

$$\vec{\mathbf{F}}_{is} = \begin{cases} \mathbf{F}_{is}, & \text{if } i, s \in \ell_0 \\ \bar{\mathbf{F}}_{is}, & \text{else} \end{cases}$$

$$\mathbf{F}_{is} = \begin{bmatrix} -\mathbf{F}_s & \mathbf{F}_s \\ 0 & \mathbf{F}_i \end{bmatrix}, \bar{\mathbf{F}}_{is} = \begin{bmatrix} -\mathbf{F}_s & \mathbf{F}_s & \mathbf{f}_s \\ 0 & \mathbf{F}_i & \mathbf{f}_i \end{bmatrix}. \qquad (5.26)$$

In order to pose the search of the Lyapunov function in a less conservative way, the S-procedure has been adopted in [66]. Along this line, the matrices $\bar{\mathbf{H}}_i = \begin{bmatrix} \mathbf{H}_i & \mathbf{h}_i \end{bmatrix}, i \in \ell$ are construced with $\mathbf{h}_i = 0$ for $i \in \ell_0$ that satisfy

$$\bar{\mathbf{H}}_i \begin{bmatrix} \mathbf{x}(t) \\ 1 \end{bmatrix} \succeq 0, \bar{\mathbf{H}}_s \begin{bmatrix} \hat{\mathbf{x}}(t) \\ 1 \end{bmatrix} \succeq 0, \quad \mathbf{x}(t) \in \mathcal{S}_i, \hat{\mathbf{x}}(t) \in \mathcal{S}_s, i, s \in \ell \quad (5.27)$$

where $\succeq$ means that each entry of the vector is nonnegative. It is evident that

$$\vec{\mathbf{H}}_{is}\vec{\eta} \succeq 0, \quad \mathbf{x}(t) \in \mathcal{S}_i, \hat{\mathbf{x}}(t) \in \mathcal{S}_s, i, s \in \ell \qquad (5.28)$$

with

$$\vec{\mathbf{H}}_{is} = \begin{cases} \mathbf{H}_{is}, & \text{if } i, s \in \ell_0 \\ \bar{\mathbf{H}}_{is}, & \text{else} \end{cases}$$

$$\mathbf{H}_{is} = \begin{bmatrix} -\mathbf{H}_s & \mathbf{H}_s \\ 0 & \mathbf{H}_i \end{bmatrix}, \bar{\mathbf{H}}_{is} = \begin{bmatrix} -\mathbf{H}_s & \mathbf{H}_s & \mathbf{h}_s \\ 0 & \mathbf{H}_i & \mathbf{h}_i \end{bmatrix}. \qquad (5.29)$$

Remark 5.3. *It is noted that a systematic procedure for constructing the continuity matrices* $\mathbf{F}_i, \mathbf{f}_i$ *and region boundary matrices* $\mathbf{H}_i, \mathbf{h}_i$ *for a given piecewise fuzzy system can be found in [66] and [65].*

The following lemma plays an essential role in deriving the main results of this chapter.

Lemma 1. *(Projection Lemma) Given a symmetric matrix* $\mathcal{W} = \mathcal{W}^T \in \mathcal{R}^{n \times n}$ *and two matrices* $\mathcal{V} \in \mathcal{R}^{m \times n}, \mathcal{U} \in \mathcal{R}^{k \times n}$, *there exists an* $\mathcal{X}$ *such that the following LMI holds*

$$\mathcal{W} + \mathcal{U}^T \mathcal{X}^T \mathcal{V} + \mathcal{V}^T \mathcal{X} \mathcal{U} < 0 \qquad (5.30)$$

if and only if

$$\begin{cases} \mathcal{U}_\perp^T \mathcal{W} \mathcal{U}_\perp < 0, \ \text{if } \mathcal{V}_\perp = 0, \ \mathcal{U}_\perp \neq 0 \\ \mathcal{V}_\perp^T \mathcal{W} \mathcal{V}_\perp < 0, \ \text{if } \mathcal{U}_\perp = 0, \ \mathcal{V}_\perp \neq 0 \\ \mathcal{U}_\perp^T \mathcal{W} \mathcal{U}_\perp < 0, \ \mathcal{V}_\perp^T \mathcal{W} \mathcal{V}_\perp < 0, \ \text{if } \mathcal{V}_\perp \neq 0, \ \mathcal{U}_\perp \neq 0 \end{cases} \qquad (5.31)$$

where $\mathcal{U}_\perp$ and $\mathcal{V}_\perp$ represent the right null spaces of $\mathcal{U}$ and $\mathcal{V}$, respectively.

Now one design scheme for the $\mathcal{L}_\infty/\mathcal{L}_2$ type of piecewise fuzzy observer-based residual generators will be addressed via piecewise quadratic Lyapunov functions in the following theorem.

Theorem 5.1. *Given nonlinear systems (5.1) and fuzzy residual generator (5.17). If there exist constants $\alpha > 0, \xi > 0$, matrices $\mathbf{T}, \mathbf{W}_{is}, \mathbf{U}_{is}, \vec{\mathcal{M}}$, $\vec{\mathcal{N}}, \vec{\mathbf{P}}_{is}, \mathbf{L}_{sm}, m \in \Gamma(s), l \in \Gamma(i), i, s \in \ell$, such that $\mathbf{W}_{is}, \mathbf{U}_{is}$ have non-negative entries and the following inequalities are feasible*

$$\vec{\mathbf{P}}_{is} - \vec{\mathbf{H}}_{is}^T \mathbf{U}_{is} \vec{\mathbf{H}}_{is} > 0, \quad i, s \in \ell \qquad (5.32)$$

$$\begin{bmatrix} -\vec{\mathbf{P}}_{is} + \vec{\mathbf{G}}_1 & \star \\ \vec{\mathbf{C}}_{ln} & -\mathbf{I} + \epsilon_3^2 \mathbf{I} \end{bmatrix} < 0, \quad n \in \Gamma(s), l \in \Gamma(i), i, s \in \ell \qquad (5.33)$$

$$\vec{\Omega}_{isnnl} < 0, \quad n \in \Gamma(s), l \in \Gamma(i), i, s \in \ell \qquad (5.34)$$

$$\vec{\Omega}_{isnml} + \vec{\Omega}_{ismnl} < 0, \quad m, n \in \Gamma(s), l \in \Gamma(i), m > n, i, s \in \ell \qquad (5.35)$$

where

$$\vec{\Omega}_{isnml} = \begin{bmatrix} -\vec{\mathcal{M}}_{is} - \vec{\mathcal{M}}_{is}^T \\ \vec{\mathbf{P}}_{is} - \vec{\mathcal{N}}_{is} + \vec{\mathbb{A}}_{snml}^T \vec{\mathcal{M}}_{is}^T \\ \mathbb{B}_{nl}^T \vec{\mathcal{M}}_{is}^T \\ \mathbf{Z}_{sm}^T \vec{\mathcal{M}}_{is}^T \end{bmatrix.$$

$$\begin{bmatrix} \star & \star & \star \\ Sym\{\vec{\mathcal{N}}_{is} \vec{\mathbb{A}}_{snml}\} + \vec{\mathbf{G}}_2 + \vec{\mathbf{H}}_{is}^T \mathbf{W}_{is} \vec{\mathbf{H}}_{is} & \star & \star \\ \mathbf{Z}_{sm}^T \vec{\mathcal{N}}_{is}^T & -\alpha^2 \mathbf{I} + \xi\lambda \mathbf{I} & \star \\ \mathbb{B}_{nl}^T \vec{\mathcal{N}}_{is}^T & 0 & -\xi\mathbf{I} \end{bmatrix}$$

$$\mathbb{A}_{snml} = \begin{bmatrix} \mathbf{A}_n - \mathbf{L}_{sm}\mathbf{C}_n & \mathbf{A}_l - \mathbf{L}_{sm}\mathbf{C}_l - \mathbf{A}_n + \mathbf{L}_{sm}\mathbf{C}_n \\ 0 & \mathbf{A}_l \end{bmatrix}$$

$$\vec{\mathbb{A}}_{snml} = \begin{bmatrix} \mathbb{A}_{snml} & \mathbf{a}_{ln} \\ 0 & 0 \end{bmatrix}, \mathbf{a}_{ln} = \begin{bmatrix} \mathbf{a}_l - \mathbf{a}_n \\ \mathbf{a}_l \end{bmatrix}$$

$$\mathbb{B}_{nl} = \begin{bmatrix} \mathbf{B}_l - \mathbf{B}_n & \mathbf{D}_l \\ \mathbf{B}_l & \mathbf{D}_l \end{bmatrix}, \bar{\mathbb{B}}_{nl} = \begin{bmatrix} \mathbb{B}_{snl} \\ \mathbf{0} \end{bmatrix}$$

$$\mathbb{C}_{ln} = \begin{bmatrix} \mathbf{C}_n & \mathbf{C}_l - \mathbf{C}_n \end{bmatrix}, \bar{\mathbb{C}}_{ln} = \begin{bmatrix} \mathbb{C}_{ln} & \mathbf{0} \end{bmatrix}$$

$$\mathbf{Z}_{sm} = \begin{bmatrix} \mathbf{I} & -\mathbf{L}_{sm} \\ \mathbf{I} & \mathbf{0} \end{bmatrix}, \bar{\mathbf{Z}}_{sm} = \begin{bmatrix} \mathbf{Z}_{sm} \\ \mathbf{0} \end{bmatrix}$$

$$\mathbf{G}_1 = \begin{bmatrix} \mathbf{0} & \mathbf{0} \\ \mathbf{0} & \epsilon_3^2 \mathbf{I}_{(k_x+k_y)\times(k_x+k_y)} \end{bmatrix}, \bar{\mathbf{G}}_1 = \begin{bmatrix} \mathbf{0} & \mathbf{0} \\ \mathbf{0} & \epsilon_3^2 \mathbf{I}_{(k_x+1)} \end{bmatrix}$$

$$\mathbf{G}_2 = \begin{bmatrix} \mathbf{0} & \mathbf{0} \\ \mathbf{0} & \xi\lambda\mathbf{I}_{k_x} \end{bmatrix}, \bar{\mathbf{G}}_2 = \begin{bmatrix} \mathbf{0} & \mathbf{0} \\ \mathbf{0} & \xi\lambda\mathbf{I}_{(k_x+1)} \end{bmatrix}, \lambda = \epsilon_1^2 + \epsilon_2^2 + \epsilon_3^2 + \epsilon_4^2$$

$$\begin{bmatrix} \vec{\mathcal{A}}_{snml} & | & \vec{\mathbb{B}}_{snl} & | & \vec{\mathbb{C}}_{ln} & | & \vec{\mathbf{G}}_1 & | & \vec{\mathbf{G}}_2 & | & \vec{\mathbf{Z}}_{sm} \end{bmatrix}$$

$$= \begin{cases} \begin{bmatrix} \mathbb{A}_{snml} & | & \mathbb{B}_{snl} & | & \mathbb{C}_{ln} & | & \mathbf{G}_1 & | & \mathbf{G}_2 & | & \mathbf{Z}_{sm} \end{bmatrix}, \text{if } i, s \in \ell_0 \\ \begin{bmatrix} \bar{\mathbb{A}}_{snml} & | & \bar{\mathbb{B}}_{snl} & | & \bar{\mathbb{C}}_{ln} & | & \bar{\mathbf{G}}_1 & | & \bar{\mathbf{G}}_2 & | & \bar{\mathbf{Z}}_{sm} \end{bmatrix}, \text{else.} \end{cases}$$

$$(5.36)$$

Then, it holds that

$$\|\mathbf{r}(\tau)\| < \alpha^2 \|\xi\|_{2,\tau}^2 + \vec{\eta}^T(0)\bar{\mathcal{P}}_0\vec{\eta}(0) \tag{5.37}$$

with $\vec{\mathcal{P}}_0 = \vec{\mathbf{P}}_{i_0 s_0}$, *where* $\theta(0) \in \mathcal{S}_{i_0}, \hat{\theta}(0) \in \mathcal{S}_{s_0}$.

Proof. Consider the following PQLFs that are continuous across the region boundaries

$$V(\vec{\eta}(t)) = \vec{\eta}^T(t)\vec{\mathbf{P}}_{is}\vec{\eta}(t), \ i, s \in \ell \tag{5.38}$$

It follows directly from Theorem 5.1 that

$$\mathbf{r}^T(t)\mathbf{r}(t) < V(\vec{\eta}(t)) \tag{5.39}$$

$$\dot{V}(\vec{\eta}(t)) - \alpha^2\xi^T(t)\xi(t) < \mathbf{0} \tag{5.40}$$

yields (7.37). Thus, in what follows, we are devoted to seek the solvability of (5.39)-(5.40). It is evident that Eqs. (5.39)-(5.40) are equivalent to

$$\vec{\eta}^T(t)\vec{\mathcal{C}}_{is}^T\vec{\mathcal{C}}_{is}\vec{\eta}(t) < \vec{\eta}^T(t)\vec{P}_{is}\vec{\eta}(t) \tag{5.41}$$

$$\vec{\eta}^T(t)\left(\vec{\mathbf{P}}_{is}\vec{\mathcal{A}}_{is} + \vec{\mathcal{A}}_{is}^T\vec{\mathbf{P}}_{is}\right)\vec{\eta}(t) + 2\vec{\eta}^T(t)\vec{\mathbf{P}}_{is}\vec{\mathcal{B}}_{is}\xi(t) - \alpha^2\xi^T(t)\xi(t) < \mathbf{0}. \tag{5.42}$$

Then based on $\mathcal{S}$-procedure, one has that that the left-hand-side (LHS) of (5.42) satisfies

$$\text{LHS(5.42)} \leq \left[\begin{array}{c} \vec{\eta}(t) \\ \xi(t) \end{array} \right]^T \left[\begin{array}{cc} \text{Sym}\left\{\vec{\mathbf{P}}_{is}\vec{\mathcal{A}}_{is}\right\} + \vec{\mathbf{H}}_{is}^T\mathbf{W}_{is}\vec{\mathbf{H}}_{is} & \star \\ \vec{\mathcal{B}}_{is}^T\vec{\mathbf{P}}_{is} & -\alpha\mathbf{I} \end{array} \right] \left[\begin{array}{c} \vec{\eta}(t) \\ \xi(t) \end{array} \right].$$

$$(5.43)$$

Thus, by Schur Complement, it is easy to see that the following inequalities imply (5.41) and (5.42)

$$\left[\begin{array}{cc} -\vec{\mathbf{P}}_{is} & \star \\ \vec{\mathbf{C}}_{is} & -\mathbf{I} \end{array} \right] < \mathbf{0} \qquad (5.44)$$

$$\left[\begin{array}{cc} \text{Sym}\left\{\vec{\mathbf{P}}_{is}\vec{\mathcal{A}}_{is}\right\} + \vec{\mathbf{H}}_{is}^T\mathbf{W}_{is}\vec{\mathbf{H}}_{is} & \star \\ \vec{\mathcal{B}}_{is}^T\vec{\mathbf{P}}_{is} & -\alpha^2\mathbf{I} \end{array} \right] < \mathbf{0}. \qquad (5.45)$$

It is noted that the Lyapunov matrices are coupling with the system matrices in (5.45). To eliminate the coupling, the Projection Lemma will be applied. To this end, we first rewrite (5.45) as follows:

$$\mathcal{U}_\perp^T \mathcal{W} \mathcal{U}_\perp < \mathbf{0} \qquad (5.46)$$

where

$$\mathcal{U}_\perp = \left[\begin{array}{cc} \vec{\mathcal{A}}_{is} & \vec{\mathcal{B}}_{is} \\ \mathbf{I} & \mathbf{0} \\ \mathbf{0} & \mathbf{I} \end{array} \right], \mathcal{W} = \left[\begin{array}{ccc} \mathbf{0} & \vec{\mathbf{P}}_{is} & \mathbf{0} \\ \vec{\mathbf{P}}_{is} & \vec{\mathbf{H}}_{is}^T\mathbf{W}_{is}\vec{\mathbf{H}}_{is} & \mathbf{0} \\ \mathbf{0} & \mathbf{0} & -\alpha^2\mathbf{I} \end{array} \right].$$

It follows directly from the explicit null space calculation that

$$\left[\begin{array}{ccc} -\mathbf{I} & \vec{\mathcal{A}}_{is} & \vec{\mathcal{B}}_{is} \end{array} \right]_\perp = \left[\begin{array}{cc} \vec{\mathcal{A}}_{is} & \vec{\mathcal{B}}_{is} \\ \mathbf{I} & \mathbf{0} \\ \mathbf{0} & \mathbf{I} \end{array} \right] = \mathcal{U}_\perp. \qquad (5.47)$$

Then, by assigning

$$\left[\begin{array}{ccc} -\mathbf{I} & \vec{\mathcal{A}}_{is} & \vec{\mathcal{B}}_{is} \end{array} \right] \to \mathcal{U}$$

$$\mathbf{I} \to \mathcal{V}, \mathbf{0} \to \mathcal{V}_\perp$$

$$\left[\begin{array}{ccc} \vec{\mathcal{M}}_{is}^T & \vec{\mathcal{N}}_{is}^T & \mathbf{0} \end{array} \right]^T \to \mathcal{X} \qquad (5.48)$$

and applying Projection lemma, it is evident that (5.46) holds if and only if the following inequality is solvable

$$
\left[
\begin{array}{ccc}
-\vec{\mathcal{M}}_{is} - \vec{\mathcal{M}}_{is}^T & \star & \star \\
\vec{\mathbf{P}}_{is} - \vec{\mathcal{N}}_{is} + \vec{\mathcal{A}}_{is}^T \vec{\mathcal{M}}_{is}^T & \mathrm{Sym}\left\{\vec{\mathcal{N}}_{is}\vec{\mathcal{A}}_{is}\right\} + \vec{\mathbf{H}}_{is}^T \mathbf{W}_{is} \vec{\mathbf{H}}_{is} & \star \\
\vec{\mathcal{B}}_{is}^T \vec{\mathcal{M}}_{is}^T & \vec{\mathcal{B}}_{is}^T \vec{\mathcal{N}}_{is}^T & -\alpha^2 \mathbf{I}
\end{array}
\right] < \mathbf{0}
$$
(5.49)

where

$$
\left\{\vec{\mathcal{M}}_{is}, \vec{\mathcal{N}}_{is}\right\} \in
\begin{cases}
\mathcal{R}^{2k_x \times 2k_x}, & \text{if } i, s \in \ell_0 \\
\mathcal{R}^{(2k_x+1)\times(2k_x+1)}, & \text{else.}
\end{cases}
$$
(5.50)

It is easy to see that the Lyapunov matrices in (5.49) have been decoupled from the system matrices. Moreover, by expanding the fuzzy-basis functions, (5.44) and (5.49) can be equivalently expressed as follows:

$$
\sum_{n\in\Gamma(s)}\sum_{l\in\Gamma(i)} \hat{\mu}_n\mu_l
\left[
\begin{array}{cc}
-\vec{\mathbf{P}}_{is} & \star \\
\vec{\mathbb{C}}_{ln} + \vec{\Delta}_{\mathbb{C}} & -\mathbf{I}
\end{array}
\right] < \mathbf{0}
$$
(5.51)

$$
\sum_{n\in\Gamma(s)}\sum_{m>n, m\in\Gamma(s)}\sum_{l\in\Gamma(i)} \hat{\mu}_n\hat{\mu}_m\mu_l \left(\vec{\Xi}_{isnml} + \vec{\Xi}_{ismnl}\right)
$$

$$
+ \sum_{n\in\Gamma(s)}\sum_{l\in\Gamma(i)} \hat{\mu}_n^2\mu_l\vec{\Xi}_{isnnl} < \mathbf{0}
$$
(5.52)

where

$$
\vec{\Xi}_{isnml} = \vec{\Pi}_{isnml} + \vec{\mathbf{F}}_{sm}\vec{\Delta}_\Pi + \vec{\Delta}_\Pi^T\vec{\mathbf{F}}_{sm}^T
$$

$$
\vec{\Pi}_{isnml} =
\left[
\begin{array}{cc}
-\vec{\mathcal{M}}_{is} - \vec{\mathcal{M}}_{is}^T & \star \\
\vec{\mathbf{P}}_{is} - \vec{\mathcal{N}}_{is} + \vec{\mathbb{A}}_{snml}^T \vec{\mathcal{M}}_{is}^T & \mathrm{Sym}\left\{\vec{\mathcal{N}}_{is}\vec{\mathbb{A}}_{snml}\right\} + \vec{\mathbf{H}}_{is}^T \mathbf{W}_{is} \vec{\mathbf{H}}_{is} \\
\vec{\mathbb{B}}_{nl}^T \vec{\mathcal{M}}_{is}^T & \vec{\mathbb{B}}_{nl}^T \vec{\mathcal{N}}_{is}^T
\end{array}
\right.
$$

$$
\left.
\begin{array}{c}
\star \\
\star \\
-\alpha^2 \mathbf{I}
\end{array}
\right]
$$

$$
\vec{\mathbf{F}}_{sm} =
\left[
\begin{array}{c}
\vec{\mathcal{M}}_{is}\vec{\mathbf{Z}}_{sm} \\
\vec{\mathcal{N}}_{is}\vec{\mathbf{Z}}_{sm} \\
\mathbf{0}
\end{array}
\right],
\Delta_{\mathbb{C}} = \left[\begin{array}{ccc} \mathbf{0} & \Delta_{\mathbb{C}}(x) & \mathbf{0} \end{array}\right]
$$

$$\Delta_\Pi = \begin{bmatrix} 0 & 0 & -\Delta_A(\mathbf{x},\mathbf{u}) & \Delta_B(\mathbf{x},\mathbf{u}) & \Delta_E(\mathbf{x},\mathbf{u}) \\ 0 & 0 & -\Delta_C(\mathbf{x}) & 0 & 0 \end{bmatrix}$$

$$\bar{\Delta}_\Pi = \begin{bmatrix} 0 & 0 & -\Delta_A(\mathbf{x},\mathbf{u}) & 0 & \Delta_B(\mathbf{x},\mathbf{u}) & \Delta_E(\mathbf{x},\mathbf{u}) \\ 0 & 0 & -\Delta_C(\mathbf{x}) & 0 & 0 & 0 \end{bmatrix}$$

$$\begin{bmatrix} \vec{\Delta}_C & | & \vec{\Delta}_\Pi \end{bmatrix} = \begin{cases} \begin{bmatrix} \Delta_{\tilde{c}} & | & \Delta_\Pi \end{bmatrix}, \text{if } i,s \in \ell_0 \\ \begin{bmatrix} \Delta_C & | & \bar{\Delta}_\Pi \end{bmatrix}, \text{ else.} \end{cases} \tag{5.53}$$

By considering the nonnegative property of the fuzzy basis function, it becomes evident that the following inequalities imply (5.51) and (5.52)

$$\begin{bmatrix} -\vec{P}_{is} & \star \\ \vec{C}_{ln} + \vec{\Delta}_C & -\mathbf{I} \end{bmatrix} < 0, \quad l \in \Gamma(i), i,s \in \ell \tag{5.54}$$

$$\vec{\Xi}_{isnnl} < 0, \quad n \in \Gamma(s), l \in \Gamma(i), i,s \in \ell \tag{5.55}$$

$$\vec{\Xi}_{isnml} + \vec{\Xi}_{ismnl} < 0, \quad m,n \in \Gamma(s), l \in \Gamma(i), m > n, i,s \in \ell. \tag{5.56}$$

For the sake of simplicity, we only focus on the proofs of (5.54) and (5.56). Note that (5.33) leads to (5.54) since

$$\begin{bmatrix} 0 & \star \\ \vec{\Delta}_C & 0 \end{bmatrix} \leq \begin{bmatrix} \vec{G}_1 & \star \\ 0 & \epsilon_3^2\mathbf{I} \end{bmatrix}, \quad i = 1, \cdots, \kappa. \tag{5.57}$$

On the other hand, it is noted for any constant $\xi > 0$, it holds that

$$(\vec{F}_{sn} + \vec{F}_{sm})\vec{\Delta}_\Pi + \vec{\Delta}_\Pi^T(\vec{F}_{sn} + \vec{F}_{sm})^T$$
$$\leq \frac{1}{2\xi}(\vec{F}_{sn} + \vec{F}_{sm})(\vec{F}_{sn} + \vec{F}_{sm})^T + 2\xi\vec{\Delta}_\Pi^T\vec{\Delta}_\Pi. \tag{5.58}$$

Moreover, it can be easily proved that for $\lambda = \epsilon_1^2 + \epsilon_2^2 + \epsilon_3^2 + \epsilon_4^2$, one obtains

$$\begin{bmatrix} -\Delta_A(\mathbf{x},\mathbf{u}) & \Delta_B(\mathbf{x},\mathbf{u}) & \Delta_E(\mathbf{x},\mathbf{u}) \\ -\Delta_C(\mathbf{x},\mathbf{u}) & 0 & 0 \end{bmatrix}^T (\star) \leq \lambda\mathbf{I}$$

$$\begin{bmatrix} -\Delta_A(\mathbf{x},\mathbf{u}) & 0 & \Delta_B(\mathbf{x},\mathbf{u}) & \Delta_E(\mathbf{x},\mathbf{u}) \\ -\Delta_C(\mathbf{x},\mathbf{u}) & 0 & 0 & 0 \end{bmatrix}^T (\star) \leq \lambda\mathbf{I} \tag{5.59}$$

which implies

$$\xi\vec{\Delta}_\Pi^T\vec{\Delta}_\Pi \leq \begin{bmatrix} 0 & 0 & 0 \\ 0 & \vec{G}_2 & \star \\ 0 & 0 & \xi\lambda\mathbf{I} \end{bmatrix}. \tag{5.60}$$

Therefore, by applying Schur Complement, it is evident that (5.35) implies (5.56). The proof is thus completed.   □

Remark 5.4. *It is noted that the following structural constraint is imposed on the multiplier $\mathcal{X}$ to derive the relax condition in (5.49)*

$$\mathcal{X} = \begin{bmatrix} \vec{\mathcal{M}}_{is}^T & \vec{\mathcal{N}}_{is}^T & 0 \end{bmatrix}^T. \tag{5.61}$$

In order to conduct (5.34) and (5.35) into a set of LMIs, the following structural constraints are imposed on $\vec{\mathcal{M}}_{is}$ and $\vec{\mathcal{N}}_{is}$, respectively

$$\mathcal{M}_{is} = \begin{bmatrix} \mathbf{M}_{s(1)} & \mathbf{M}_{is(1)} \\ 0 & \mathbf{M}_{is(2)} \end{bmatrix}, \mathcal{N}_{is} = \begin{bmatrix} \mathbf{M}_{s(1)} & \mathbf{N}_{is(1)} \\ 0 & \mathbf{N}_{is(2)} \end{bmatrix}$$

$$\vec{\mathcal{M}}_{is} = \begin{bmatrix} \mathbf{M}_{s(1)} & \mathbf{M}_{is(1)} & \mathbf{m}_{is(1)} \\ 0 & \mathbf{M}_{is(2)} & \mathbf{m}_{is(2)} \\ 0 & \mathbf{m}_{is(3)} & m_{is(4)} \end{bmatrix}$$

$$\vec{\mathcal{N}}_{is} = \begin{bmatrix} \mathbf{M}_{s(1)} & \mathbf{N}_{is(1)} & \mathbf{n}_{is(1)} \\ 0 & \mathbf{N}_{is(2)} & \mathbf{n}_{is(2)} \\ 0 & \mathbf{n}_{is(3)} & n_{is(4)} \end{bmatrix}$$

$$\begin{bmatrix} \vec{\mathcal{M}}_{is} & | & \vec{\mathcal{N}}_{is} \end{bmatrix} = \begin{cases} \begin{bmatrix} \mathcal{M}_{is} & | & \mathcal{N}_{is} \end{bmatrix}, \text{if } i, s \in \ell_0 \\ \begin{bmatrix} \vec{\mathcal{M}}_{is} & | & \vec{\mathcal{N}}_{is} \end{bmatrix}, \text{else.} \end{cases} \tag{5.62}$$

Thus, by setting $\mathbf{Q}_{sn} = \mathbf{M}_{s(11)}\mathbf{L}_{sn}$, (5.34) and (5.35) can be solved by applying the convex optimization algorithm.

It has been observed that by Projection lemma and introducing the up-triangular structural slack matrices $\vec{\mathcal{M}}_{is}, \vec{\mathcal{N}}_{is}$, the Lypaunov matrices $\vec{\mathbf{P}}_{is}$ in (5.49) have been separated from the system matrices. However, the structural constraints imposed on the slack matrices would lead to some degree of design conservatism. To circumvent this problem, the following design approach is further proposed.

Theorem 5.2. *Given nonlinear systems (5.1) and fuzzy residual generator (5.17). If there exist constants $\alpha > 0, \bar{\xi} > 0, \varsigma > 0$, matrices $\bar{\mathbf{T}}, \mathbf{W}_{is}, \bar{\mathbf{U}}_{is}, \vec{\mathcal{M}}, \vec{\mathcal{N}}, \vec{\mathcal{Q}}_{is}, \vec{\mathcal{P}}_{is}, \mathbf{L}_{sm}, m \in \Gamma(s), l \in \Gamma(i), i, s \in \ell$, such that $\mathbf{W}_{is}, \bar{\mathbf{U}}_{is}$ have nonnegative entries and the following inequalities are feasible*

$$\vec{\mathcal{P}}_{is} - \vec{\mathbf{E}}_{is}^T \bar{\mathbf{U}}_{is} \vec{\mathbf{E}}_{is} > 0, \quad i, s \in \ell \tag{5.63}$$

$$\begin{bmatrix} -\vec{\mathcal{P}}_{is} + \varrho\vec{\mathbf{G}}_1 & \star \\ \varrho\vec{\mathbf{C}}_{ln} & -\varrho\mathbf{I} + \varrho\epsilon_3^2\mathbf{I} \end{bmatrix} < 0, \quad n \in \Gamma(s), l \in \Gamma(i), i, s \in \ell \quad (5.64)$$

$$\vec{\Psi}_{isnnl} < 0, \quad n \in \Gamma(s), l \in \Gamma(i), i, s \in \ell \tag{5.65}$$

$$\vec{\Psi}_{isnml} + \vec{\Psi}_{ismnl} < 0, \quad m, n \in \Gamma(s), l \in \Gamma(i), m > n, i, s \in \ell \tag{5.66}$$

$$\vec{\mathcal{P}}_{is}\vec{\mathcal{Q}}_{is} = \mathbf{I}, \quad \varrho_\varsigma = 1, \quad i, s \in \ell \tag{5.67}$$

where

$$\vec{\Psi}_{isnml} = \begin{bmatrix} \vec{\mathbf{E}}_{is}^T W_{is}\vec{\mathbf{E}}_{is} - \vec{\mathcal{P}}_{is} + \varsigma\vec{\mathbf{G}}_2 & \star & \star \\ 0 & -\alpha^2\mathbf{I} + \bar{\xi}\lambda\mathbf{I} & \star \\ I + \varsigma\vec{\mathbb{A}}_{snml} & \varsigma\mathbb{B}_{nl} & -\mathcal{Q}_{is} + \bar{\xi}\lambda\mathbf{I} \\ 0 & 0 & \varsigma\mathbf{Z}_{sm} \end{bmatrix}$$
$$\begin{matrix} \star \\ \star \\ \star \\ -\bar{\xi}\mathbf{I} \end{matrix}$$

Then, it holds that

$$\|\mathbf{r}(\tau)\| < \alpha \|\xi\|_{2,\tau}^2 + \vec{\eta}^T(0)\vec{\mathcal{P}}_0\vec{\eta}(0) \tag{5.68}$$

with $\vec{\mathcal{P}}_0 = \vec{\mathcal{P}}_{i_0s_0}$, *where* $\theta(0) \in \mathcal{S}_{i_0}, \hat{\theta}(0) \in \mathcal{S}_{s_0}$.

Proof. It is easy to see that (5.45) can be equivalently rewritten as

$$\begin{bmatrix} \vec{\mathbf{E}}_{is}^T\mathbf{W}_{is}\vec{\mathbf{E}}_{is} & \star \\ 0 & -\alpha\mathbf{I} \end{bmatrix} + \mathrm{Sym}\left\{ \begin{bmatrix} \vec{\mathbf{P}}_{is} \\ 0 \end{bmatrix} \begin{bmatrix} \vec{\mathcal{A}}_{is} & \vec{\mathcal{B}}_{is} \end{bmatrix} \right\} < 0, \quad i, s \in \ell. \tag{5.69}$$

It is noted that

$$\begin{bmatrix} \vec{\mathbf{P}}_{is} \\ 0 \end{bmatrix} \begin{bmatrix} \vec{\mathcal{A}}_{is} & \vec{\mathcal{B}}_{is} \end{bmatrix} = \begin{bmatrix} \vec{\mathbf{P}}_{is} & 0 \\ 0 & \mathbf{I} \end{bmatrix} \begin{bmatrix} \vec{\mathcal{A}}_{is} & \vec{\mathcal{B}}_{is} \\ 0 & 0 \end{bmatrix}$$

$$\begin{bmatrix} \vec{\mathcal{A}}_{is} & \vec{\mathcal{B}}_{is} \\ 0 & 0 \end{bmatrix}^T \begin{bmatrix} \vec{\mathbf{P}}_{is} & 0 \\ 0 & \mathbf{I} \end{bmatrix} \begin{bmatrix} \vec{\mathcal{A}}_{is} & \vec{\mathcal{B}}_{is} \\ 0 & 0 \end{bmatrix} > 0. \tag{5.70}$$

Then, along the lines of [113] and [118], it is evident that there exists a sufficiently small positive constant ς such that the following inequality implies (5.69)

$$\varsigma \begin{bmatrix} \vec{\mathcal{A}}_{is} & \vec{\mathcal{B}}_{is} \\ 0 & 0 \end{bmatrix}^T \begin{bmatrix} \vec{\mathbf{P}}_{is} & 0 \\ 0 & \mathbf{I} \end{bmatrix} \begin{bmatrix} \vec{\mathcal{A}}_{is} & \vec{\mathcal{B}}_{is} \\ 0 & 0 \end{bmatrix} + \text{LHS}(5.69) < 0. \quad (5.71)$$

By Schur complement, one obtains that

$$\begin{bmatrix} \vec{\mathbf{E}}_{is}^T \mathbf{W}_{is} \vec{\mathbf{E}}_{is} - \varsigma^{-1} \vec{\mathbf{P}}_{is} & \star & \star \\ 0 & -\alpha^2 \mathbf{I} & \star \\ \mathbf{I} + \varsigma \vec{\mathcal{A}}_{is} & \varsigma \vec{\mathcal{B}}_{is} & -\varsigma \vec{\mathbf{P}}_{is}^{-1} \end{bmatrix} < 0. \quad (5.72)$$

Denote $\vec{\mathcal{P}}_{is} = \varsigma^{-1} \vec{\mathbf{P}}_{is}, \bar{\mathbf{T}} = \varsigma^{-1} \mathbf{T}, \vec{\mathcal{Q}}_{is} = \varsigma \vec{\mathbf{P}}_{is}^{-1}, \bar{\mathbf{U}}_{is} = \varsigma^{-1} \mathbf{U}_{is}, \bar{\xi} = \varsigma \xi$. Similar to the proof of Theorem 3, by extracting the fuzzy-basis functions and convexifying the uncertainties, it is easy to see that (5.44) and (5.72) are fulfilled provided that (5.63)-(5.67) are feasible. The proof is thus completed. □

Remark 5.5. *It is worth mentioning that by adopting cone complementarity method [50], the nonconvex problem proposed in Theorem 5.2 can be converted to the following minimization problem*

$$\text{Minimize Trace} \left(\sum_{i,s \in \ell} \vec{\mathcal{P}}_{is} \vec{\mathcal{Q}}_{is} + \varrho \varsigma \right)$$

subject to (5.63) − (5.67), and

$$\vec{\mathcal{Q}}_{is} > 0, \begin{bmatrix} \vec{\mathcal{P}}_{is} & \mathbf{I} \\ \mathbf{I} & \vec{\mathcal{Q}}_{is} \end{bmatrix} > 0, \begin{bmatrix} \varrho & 1 \\ 1 & \varsigma \end{bmatrix} > 0, i, s \in \ell. \quad (5.73)$$

5.2.3 $\mathcal{L}_\infty/\mathcal{L}_2$ Type of Fuzzy Observer-based FD Systems

Theorem 5.1 provides an algorithm for the design of an $\mathcal{L}_\infty/\mathcal{L}_2$ type of piecewise fuzzy observer-based residual generator. Based on the residual generator, by (i) defining the residual evaluation function with a finite time evaluation window as

$$J(\mathbf{r}) = ||\mathbf{r}_\tau||_2^2 = \int_{t_0}^{t_0+\tau} \mathbf{r}^T(t)\mathbf{r}(t)dt \quad (5.74)$$

(ii) determining the dynamic threshold

$$J_{\text{th}} = \alpha^2 \left\| \mathbf{u}_\tau \right\|_2^2 + \alpha^2 \sigma_d^2 + \bar{\gamma}_o$$

$$\bar{\gamma}_o = \sup_{\mathbf{x}(0), \hat{\mathbf{x}}(0)} \{\gamma_0\}, \gamma_o = \vec{\eta}^T(0)\vec{P}_0\vec{\eta}(0) \tag{5.75}$$

with $\vec{P}_0 = \vec{P}_{i_0 s_0}$, where $\theta(0) \in \mathcal{S}_{i_0}, \hat{\theta}(0) \in \mathcal{S}_{s_0}$, and (iii) setting the decision logic, an $\mathcal{L}_\infty/\mathcal{L}_2$ type of fuzzy observer-based FD system can be realized. The on-line algorithm of $\mathcal{L}_\infty/\mathcal{L}_2$ type of fuzzy observer-based FD method for nonlinear systems is summarized in Algorithm 2.

Algorithm 2 On-line early fault detection algorithm for nonlinear processes

---

1: Run the residual generator (5.17)
2: Run the evaluator (5.74)
3: Set the adaptive threshold

$$J_{\text{th}} = \alpha \left\| \mathbf{u}_\tau \right\|_2^2 + \alpha \sigma_d^2 + \bar{\gamma}_o \tag{5.76}$$

4: Run the decision logic

$$\begin{cases} J(\mathbf{r}) = \left\| \mathbf{r}_\tau \right\|_2^2 > J_{\text{th}} \implies \text{faulty} \\ J(\mathbf{r}) = \left\| \mathbf{r}_\tau \right\|_2^2 \leq J_{\text{th}} \implies \text{fault-free} \end{cases} \tag{5.77}$$

5.3 A Numerical Example

Consider the following T-S fuzzy system with three fuzzy rules
    Plant rule $\Re^i$: IF $x_1(t)$ is N_1^i, THEN

$$\begin{cases} \dot{\mathbf{x}} = \mathbf{A}_i \mathbf{x} + \mathbf{B}_i u + \mathbf{E}_i d + \mathbf{a}_i + \Delta_{\mathbf{A}}(\mathbf{x}, u)\mathbf{x} + \Delta_{\mathbf{B}}(\mathbf{x}, u)\mathbf{u} + \Delta_{\mathbf{E}}(\mathbf{x}, u)\mathbf{d} \\ \mathbf{y} = \mathbf{C}_i \mathbf{x} + \Delta_{\mathbf{C}}(\mathbf{x})\mathbf{x}, \quad i \in \{1, 2, 3\} \end{cases}$$

where

$$\mathbf{A}_1 = \begin{bmatrix} -2 & 1 \\ -2.1 & -1 \end{bmatrix}, \mathbf{B}_1 = \begin{bmatrix} 0.1 \\ 0.1 \end{bmatrix}, \mathbf{E}_1 = \begin{bmatrix} 0.1 \\ 0 \end{bmatrix}$$

$$\mathbf{A}_2 = \begin{bmatrix} -3 & 1 \\ -2 & -1 \end{bmatrix}, \mathbf{B}_2 = \begin{bmatrix} 0.2 \\ 0.1 \end{bmatrix}, \mathbf{E}_2 = \begin{bmatrix} 0.1 \\ 0 \end{bmatrix}$$

$$\mathbf{A}_3 = \begin{bmatrix} -2 & 1 \\ -2 & -1 \end{bmatrix}, \mathbf{B}_3 = \begin{bmatrix} 0.2 \\ 0.1 \end{bmatrix}, \mathbf{E}_3 = \begin{bmatrix} 0.1 \\ 0 \end{bmatrix}$$

$$\mathbf{a}_1 = \begin{bmatrix} -1 \\ 0 \end{bmatrix}, \mathbf{a}_2 = \begin{bmatrix} 0 \\ 0 \end{bmatrix}, \mathbf{a}_3 = \begin{bmatrix} 0 \\ -1 \end{bmatrix}$$

$$\mathbf{C}_1 = \begin{bmatrix} 0.15 & 0.26 \end{bmatrix}, \mathbf{C}_2 = \begin{bmatrix} 0.25 & 0.26 \end{bmatrix}, \mathbf{C}_3 = \begin{bmatrix} 0.5 & 0.26 \end{bmatrix}.$$

It is assumed that $\|\Delta_\mathbf{A}(\mathbf{x}, u)\| \leq 0.03$, $\|\Delta_\mathbf{B}(\mathbf{x}, u)\| \leq 0.01$, $\|\Delta_\mathbf{E}(\mathbf{x}, u)\| \leq 0.01$ and $\|\Delta_\mathbf{C}(\mathbf{x})\| \leq 0.01$. The membership functions for premise variable $\theta(t) = x_1(t)$ are given in Fig. 5.1. Based on the partition method proposed in Section 5.2, the space for the premise variable can be divided into three regions, which are given by

$$\mathcal{S}_1 := \{\theta| -l_1 < \theta \leq -l_2\}$$
$$\mathcal{S}_2 := \{\theta| -l_2 < \theta \leq l_3\}$$
$$\mathcal{S}_3 := \{\theta|l_3 < \theta \leq l_4\}$$

where $l_1 = 24$, $l_2 = 12$, $l_3 = 12$, $l_4 = 24$. Along the lines of [66], the constraint matrices can be constructed as follows:

$$\vec{\mathbf{F}}_1 = \begin{bmatrix} -1 & 0 & -l_2 \\ 0 & 0 & 0 \\ 1 & 0 & 0 \\ 0 & 1 & 0 \end{bmatrix}, \mathbf{F}_2 = \begin{bmatrix} 0 & 0 \\ 0 & 0 \\ 1 & 0 \\ 0 & 1 \end{bmatrix}, \vec{\mathbf{F}}_3 = \begin{bmatrix} 0 & 0 & 0 \\ 1 & 0 & -l_3 \\ 1 & 0 & 0 \\ 0 & 1 & 0 \end{bmatrix}$$

$$\vec{\mathbf{H}}_1 = \begin{bmatrix} 1 & 0 & l_1 \\ -1 & 0 & -l_2 \end{bmatrix}, \mathbf{H}_2 = \begin{bmatrix} 0 & 0 \\ 0 & 0 \end{bmatrix}, \vec{\mathbf{H}}_3 = \begin{bmatrix} 1 & 0 & -l_1 \\ -1 & 0 & l_4 \end{bmatrix}.$$

By following the algorithm proposed in Theorem 5.2 and adopting the cone complementarity method in (5.73), one obtains that $\alpha = 0.5$ and

$$\mathbf{L}_{11} = \begin{bmatrix} 0.0780 \\ 0.1153 \end{bmatrix}, \mathbf{L}_{12} = \begin{bmatrix} 0.1025 \\ 0.1153 \end{bmatrix}, \mathbf{L}_{22} = \begin{bmatrix} 0.1114 \\ 0.1156 \end{bmatrix}$$

$$\mathbf{L}_{32} = \begin{bmatrix} 0.1387 \\ 0.1154 \end{bmatrix}, \mathbf{L}_{33} = \begin{bmatrix} 0.1939 \\ 0.1156 \end{bmatrix}.$$

5.4 Concluding Remarks

The major objective of this chapter is to study the analysis integrated design issues of an early and real-time ($\mathcal{L}_\infty/\mathcal{L}_2$ type of) observer-based FD

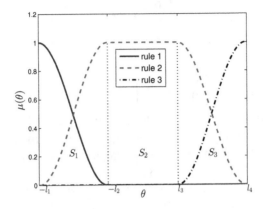

Figure 5.1: Membership functions for fuzzy modelling

systems for nonlinear processes in the presence of external disturbances. To gain a deeper insight into this type of FD framework, the existence and design conditions have been discussed first. Then, a T-S fuzzy-model-based design scheme of this $\mathcal{L}_\infty/\mathcal{L}_2$ type of observer-based FD systems has been developed via piecewise Lyapunov functions. In particular, this scheme can be applied to the case that the premise variables of the T-S fuzzy models of the plant is not available. This work is strongly motivated by the increasing requirement on safety and reliability of complex industrial processes.

6 Design of Weighted Fuzzy Observer-based FD Systems

In recent years, gain-scheduled approaches have been extensively investigated for the analysis and design of controllers and observers for nonlinear systems or parameter dependent dynamic processes, such as Markov jump systems, LPV systems [75, 119, 121, 125]. By establishing the relationship between a nonlinear system and a family of linear systems, the design of nonlinear controllers and observers can be realized via a local depending gain-scheduled table. Owing to this significant role in solving nonlinear design issues, we are more interested in studying fuzzy observer-based FD schemes for nonlinear systems in this gain-scheduling manner.

Inspired by the above observations, the major objective of this chapter is on the analysis and integrated design of $\mathcal{L}_2$ observer-based FD systems for a general type of discrete-time nonlinear processes. To this end, the existence condition of the nonlinear FD systems is studied first. Then, the major attention is paid to the design approach by solving the proposed existence condition with the help of T-S fuzzy dynamic modelling technique and the convex optimization algorithm. For the purpose of improving the FD performance of the conventional fuzzy observer-based FD schemes, a weighted piecewise-fuzzy observer-based residual generator is proposed. The basic idea consists in weighting the residual signals of each local region individually, instead of constantly, by means of weighting factors. Moreover, an integration of residual evaluation and threshold computation procedures into the piecewise-fuzzy FD scheme is achieved in a way such that the overall FD performance will be optimized. In the end, the $\mathcal{L}_2$ robust piecewise-fuzzy observer-based FD systems are investigated for nonlinear processes with external disturbances.

This work is strongly motivated by the fact that one of the essential ideas of T-S fuzzy-model-based control and observation schemes is to design a controller or an observer such that the overall performance would be satisfied for the worst case amongst all local behaviors. However, different from linear systems with unified dynamics over the whole working range, the local behavior of nonlinear systems can be considerably different. Thus, to reach a high FD performance, all available information of each local region should be taken into account in the design scheme of FD system. From this viewpoint, we are convinced that individually addressing the system features and making use of information provided by each local region would improve the FD performance of the overall system.

6.1 Preliminaries and Problem Formulation

In this chapter, the configuration of $\mathcal{L}_2$ observer-based FD systems for discrete-time nonlinear processes is introduced, and then the FD performance is studied.

6.1.1 Configuration of Discrete-Time $\mathcal{L}_2$ Observer-based FD Systems

Consider the discrete-time nonlinear processes of the following general form

$$\begin{cases} \mathbf{x}(k+1) = \mathbf{f}\left(\mathbf{x}(k), \mathbf{u}(k)\right) \\ \mathbf{y}(k) = \mathbf{h}\left(\mathbf{x}(k), \mathbf{u}(k)\right) \end{cases} \tag{6.1}$$

where $\mathbf{x}(k) \in \mathcal{R}^{k_x}$ is the system state; $\mathbf{y}(k) \in \mathcal{R}^{k_y}$ is the measurable output; $\mathbf{u}(k) \in \mathcal{R}^{k_u}$ is the controlled input. $\mathbf{f} : \mathcal{R}^{k_x} \times \mathcal{R}^{k_u} \to \mathcal{R}^{k_x}$ and $\mathbf{h} : \mathcal{R}^{k_x} \times \mathcal{R}^{k_u} \to \mathcal{R}^{k_y}$ are continuous nonlinear functions.

Without loss of generality, the influence of fault on nonlinear system (6.1) can be modelled as

$$\begin{cases} \mathbf{x}(k+1) = \tilde{\mathbf{f}}\left(\mathbf{x}(k), \mathbf{u}(k), \mathbf{w}(k)\right) \\ \mathbf{y}(k) = \tilde{\mathbf{h}}\left(\mathbf{x}(k), \mathbf{u}(k), \mathbf{w}(k)\right) \end{cases} \tag{6.2}$$

where $\mathbf{w} \in \mathcal{R}^{k_w}$ denotes the fault vector. $\tilde{\mathbf{f}} : \mathcal{R}^{k_x} \times \mathcal{R}^{k_u} \times \mathcal{R}^{k_w} \to \mathcal{R}^{k_x}$ and $\tilde{\mathbf{h}} : \mathcal{R}^{k_x} \times \mathcal{R}^{k_u} \times \mathcal{R}^{k_w} \to \mathcal{R}^{k_y}$ are continuous nonlinear functions. In addition, it holds that $\tilde{\mathbf{f}}\left(\mathbf{x}(k), \mathbf{u}(k), \mathbf{0}\right) = \mathbf{f}\left(\mathbf{x}(k), \mathbf{u}(k)\right), \tilde{\mathbf{h}}\left(\mathbf{x}(k), \mathbf{u}(k), \mathbf{0}\right) = \mathbf{h}\left(\mathbf{x}(k), \mathbf{u}(k)\right)$.

The following type of observer-based residual generators is adopted for the FD purpose

$$\begin{cases} \hat{\mathbf{x}}(k+1) = \varphi(\hat{\mathbf{x}}(k), \mathbf{u}(k), \mathbf{y}(k)) \\ \mathbf{r}(k) = \mathbf{y}(k) - \mathbf{h}(\hat{\mathbf{x}}(k), \mathbf{u}(k)) \end{cases} \tag{6.3}$$

where $\varphi : \mathcal{R}^{k_x} \times \mathcal{R}^{k_u} \times \mathcal{R}^{k_y} \to \mathcal{R}^{k_x}$ is continuous nonlinear function with appropriate dimensions. It should be pointed out that the residual signal $\mathbf{r}(k)$ generated by (6.3) satisfies (i) $\forall \hat{\mathbf{x}}(0) = \mathbf{x}(0)$ and $\mathbf{u}(k)$, $\mathbf{r}(k) \equiv \mathbf{0}$ in fault-free case; (ii) for some $\mathbf{w} \neq \mathbf{0}$ in (6.2), $\mathbf{r}(k) \neq \mathbf{0}$.

Throughout of this chapter, we consider the evaluation function as an $\mathcal{L}_2$ norm of $\mathbf{r}(k)$ with an evaluation window $[0, \tau]$

$$J(\mathbf{r}) = \sum_{k=0}^{\tau} \varphi_1\left(\|\mathbf{r}(k)\|\right) \tag{6.4}$$

where $\varphi_1\left(\cdot\right)$ belongs to the class of $\mathcal{K}$-function.

Moreover, by defining the maximum influence of known or unknown inputs and initial conditions on residual signal in fault-free case as the threshold

$$J_{\text{th}} = \sup_{\mathbf{x}(0), \hat{\mathbf{x}}(0), \mathbf{w}=\mathbf{0}} J(\mathbf{r}) \tag{6.5}$$

it allows then the application of the following decision logic

$$\begin{cases} J(\mathbf{r}) > J_{\text{th}} \implies \text{faulty} \\ J(\mathbf{r}) \leq J_{\text{th}} \implies \text{fault-free} \end{cases} \tag{6.6}$$

for a successful $\mathcal{L}_2$ observer-based FD for nonlinear discrete-time systems.

6.1.2 The Existence Condition of Discrete-Time $\mathcal{L}_2$ Observer-based FD Systems

In this section, the existence condition for the aforementioned FD systems for discrete-time nonlinear processes (6.1) will be studied.

To this end, a more general type of residual generators is introduced first

$$\begin{cases} \hat{\mathbf{x}}(k+1) = \varphi(\hat{\mathbf{x}}(k), \mathbf{u}(k), \mathbf{y}(k)) \\ \mathbf{r}(k) = \psi(\mathbf{y}(k) - \mathbf{h}(\hat{\mathbf{x}}(k), \mathbf{u}(k))) \end{cases} \tag{6.7}$$

where $\mathbf{r}(k) \in \mathcal{R}^{k_r}$ is residual signal. $\psi : \mathcal{R}^{k_y} \to \mathcal{R}^{k_r}$ is nonlinear function. It is noteworthy that $\mathbf{r}(k) = \psi(\mathbf{y}(k) - \mathbf{h}(\hat{\mathbf{x}}(k), \mathbf{u}(k)))$, instead of $\mathbf{r}(k) = \mathbf{y}(k) - \mathbf{h}(\hat{\mathbf{x}}(k), \mathbf{u}(k))$, is implemented here for the generation of residual signal. This is motivated by the parametrization form of nonlinear observer-based residual generators and its application in optimizing residual evaluation and threshold setting [1]. That is to say, this generalization form provides us the potential to improve the FD performance, which will be further addressed in terms of weighted piecewise-fuzzy observer-based residual generators in the forthcoming sections.

For the FD system design purpose, the $\mathcal{L}_2$ re-constructible concept is extended to discrete-time nonlinear systems first.

Definition 6.1. *System (6.1) is said to be $\mathcal{L}_2$ re-constructible if there exists a nonlinear system (6.7), and positive constant δ such that $\forall \mathbf{x}(0), \hat{\mathbf{x}}(0) \in \mathcal{B}_\delta$*

$$\sum_{k=0}^{\tau} \varphi_1(\|\mathbf{r}(k)\|) \leq \sum_{k=0}^{\tau} \varphi_2(\|\mathbf{u}(k)\|) + \gamma_0(\mathbf{x}(0), \hat{\mathbf{x}}(0)) \qquad (6.8)$$

where $\varphi_1(\cdot) \in \mathcal{K}, \varphi_2(\cdot) \in \mathcal{K}_\infty$ and $\gamma_0(\cdot) \geq 0$ is a (finite) constant for given $\mathbf{x}(0), \hat{\mathbf{x}}(0)$.

Then, a sufficient condition for the $\mathcal{L}_2$ re-constructability is introduced in the following theorem, which also serves as a sufficient condition for the existence of an $\mathcal{L}_2$ observer-based FD system and the threshold setting.

Theorem 6.1. *Given system (6.1), if there exist (i) functions $\phi : \mathcal{R}^{k_x} \times \mathcal{R}^{k_u} \times \mathcal{R}^{k_y} \to \mathcal{R}^{k_x}$ and $\psi : \mathcal{R}^{k_y} \to \mathcal{R}^{k_r}$; (ii) functions $V : \mathcal{R}^{k_x} \times \mathcal{R}^{k_x} \to \mathcal{R}^+, \varphi_1(\cdot) \in \mathcal{K}, \varphi_2(\cdot) \in \mathcal{K}_\infty$ and positive constant δ such that $\forall \mathbf{x}(0), \hat{\mathbf{x}}(0) \in \mathcal{B}_\delta$*

$$V(\mathbf{f}(\mathbf{x}, \mathbf{u}), \phi(\hat{\mathbf{x}}, \mathbf{u}, \mathbf{h}(\mathbf{x}, \mathbf{u}))) - V(\mathbf{x}, \hat{\mathbf{x}}) \leq -\varphi_1(\|\mathbf{r}\|) + \varphi_2(\|\mathbf{u}\|)$$
$$\mathbf{r} = \psi(\mathbf{y}(k) - \mathbf{h}(\hat{\mathbf{x}}(k), \mathbf{u}(k))) \qquad (6.9)$$

then system (6.1) is $\mathcal{L}_2$ re-constructible. As a result, an $\mathcal{L}_2$ observer-based FD system can be realized by (i) constructing residual generator (6.7) (ii) defining the residual evaluation function as

$$J(\mathbf{r}) = \sum_{k=0}^{\tau} \varphi_1(\|\mathbf{r}(k)\|) \qquad (6.10)$$

and (iii) setting the threshold

$$J_{th} = \sum_{k=0}^{\tau} \varphi_2 \left(\|\mathbf{u}(k)\| \right) + \bar{\gamma}_0$$

$$\bar{\gamma}_0 = \sup_{\mathbf{x}(0), \hat{\mathbf{x}}(0)} V \left(\mathbf{x}(0), \hat{\mathbf{x}}(0) \right). \tag{6.11}$$

Proof. It is obvious from (6.9) that

$$V(\mathbf{x}(k+1), \hat{\mathbf{x}}(k+1)) - V(\mathbf{x}(k), \hat{\mathbf{x}}(k))$$
$$\leq -\varphi_1(\|\mathbf{r}(k)\|) + \varphi_2(\|\mathbf{u}(k)\|) \tag{6.12}$$

which implies

$$V(\mathbf{x}(\tau+1), \hat{\mathbf{x}}(\tau+1)) - V(\mathbf{x}(0), \hat{\mathbf{x}}(0))$$
$$\leq -\sum_{k=0}^{\tau} \varphi_1(\|\mathbf{r}(k)\|) + \sum_{k=0}^{\tau} \varphi_2(\|\mathbf{u}(k)\|) \tag{6.13}$$

As a consequent, we have

$$\sum_{k=0}^{\tau} \varphi_1(\|\mathbf{r}(k)\|) \leq \sum_{k=0}^{\tau} \varphi_2(\|\mathbf{u}(k)\|) + V(\mathbf{x}(0), \hat{\mathbf{x}}(0)) \tag{6.14}$$

which completes the proof. □

It is worth mentioning, if (6.9) is solvable, the $\mathcal{L}_2$ observer-based FD system proposed in Theorem 6.1 can be applied. Thus, it is of reasonable interest to seek the solution of (6.9). For this purpose, we will investigate the $\mathcal{L}_2$ observer-based FD approach in the subsequent sections by applying the well-developed fuzzy dynamic modelling technique as the solution tool.

6.1.3 FD Performance

There are different ways to evaluate the performance of an FD system [94, 25, 30]. In our study, we follow the evaluation scheme proposed in [30] and choose the so-called fault detectability as the key indicator. As described below, the performance of an FD system will be evaluated in terms of the fault detectability. Suppose that a threshold J_{th} is given based on (6.5). Denote the set of all detectable faults by

$$F_{(\mathbf{r},J,J_{\text{th}})} = \{\mathbf{w}|\mathbf{w} \neq \mathbf{0} \text{ and } J(\mathbf{r}) > J_{\text{th}}\}. \qquad (6.15)$$

It is evident that the domain of $F_{(\mathbf{r},J,J_{\text{th}})}$ characterizes the fault detectability of the FD systems. In fact, given two residual generators with residuals $\mathbf{r}_1$ and $\mathbf{r}_2$ and the same threshold J_{th}, it is reasonable to say that the residual $\mathbf{r}_2$ is more sensitive to the fault than $\mathbf{r}_1$ if

$$F_{(\mathbf{r}_1,J(\mathbf{r}_1),J_{th})} \subseteq F_{(\mathbf{r}_2,J(\mathbf{r}_2),J_{th})}. \qquad (6.16)$$

It is noted that the above norm-based threshold computation scheme (6.5) means a worst-case handling, i.e., J_{th} would cover all possible changes in the residual signal caused by unknown or known inputs and initial conditions. In this sense, only knowledge of the possible maximum influence of unknown or known inputs and initial conditions on residual signal is taken into consideration. As a consequent, the fault detectability may suffer from such a conservative setting.

6.1.4 Problem Formulation

The major objective of this chapter is to address the design approach of $\mathcal{L}_2$ observer-based FD systems for nonlinear processes (6.1) in the integrated context. To be specific, a design scheme is derived by solving the proposed condition with the aid of fuzzy dynamic modelling technique. In particular, a weighted piecewise-fuzzy observer-based residual generator is proposed based on the knowledge of each local region. For this purpose, the weighting factors will be introduced, which weight the residual signal of each local region individually. As a result, the fault detectability of the proposed approach will be improved compared to the conventional fuzzy FD methods. Moreover, the extension to FD issues of nonlinear systems with external disturbances will be studied.

6.2 $\mathcal{L}_2$ Piecewise-Fuzzy Observer-based FD Systems

In the above section, the existence condition of $\mathcal{L}_2$ observer-based FD systems has been investigated. In this section, the design issues will be addressed by applying piecewise-fuzzy dynamic modelling technique.

6.2.1 Piecewise-Fuzzy Dynamic Modelling

It is known that by including the input signals into the premise variables
, the following class of generalized T-S fuzzy models can be employed
to approximate a general type of discrete-time nonlinear systems (6.1)
[155, 48]:

Plant rule $\Re^l$: IF $\theta_1(k)$ is N_1^l and $\theta_2(k)$ is N_2^l and $\cdots$ and $\theta_p(k)$ is N_p^l,
THEN

$$\begin{cases} \mathbf{x}(k+1) = \mathbf{A}_l\mathbf{x}(k) + \mathbf{B}_l\mathbf{u}(k) + \Delta_{\mathbf{A}}(\mathbf{x},\mathbf{u})\mathbf{x} + \Delta_{\mathbf{B}}(\mathbf{x},\mathbf{u})\mathbf{u} \\ \mathbf{y}(k) = \mathbf{C}_l\mathbf{x}(k) + \mathbf{D}_l\mathbf{u}(k) + \Delta_{\mathbf{C}}(\mathbf{x},\mathbf{u})\mathbf{x} + \Delta_{\mathbf{D}}(\mathbf{x},\mathbf{u})\mathbf{u} \\ \qquad\qquad\qquad l \in \{1,2,\cdots,\kappa\} \end{cases} \quad (6.17)$$

where $\theta(k) = [\theta_1(k) \ \cdots \ \theta_p(k)]$ are the premise variables; $N_j^l(j=1,\cdots,p)$
represents the fuzzy sets; $\Re^l$ denotes the lth fuzzy inference rule; κ
indicates the number of inference rules; $\mathbf{A}_l$, $\mathbf{B}_l$, $\mathbf{C}_l$ and $\mathbf{D}_l$ are known
system matrices with appropriate dimensions; $\mathbf{x}(k)$, $\mathbf{u}(k)$ and $\mathbf{y}(k)$ denote
the process state, input and output variables, respectively. As discussed
in Chapter 4, for any given positive constants $\epsilon_1, \epsilon_2, \epsilon_3$ and ϵ_4, there exists
T-S fuzzy dynamic model (6.19) for nonlinear systems (6.1) such that

$$\|\Delta_{\mathbf{A}}(\mathbf{x},\mathbf{u})\| \leq \epsilon_1, \ \|\Delta_{\mathbf{B}}(\mathbf{x},\mathbf{u})\| \leq \epsilon_2$$
$$\|\Delta_{\mathbf{C}}(\mathbf{x},\mathbf{u})\| \leq \epsilon_3, \ \|\Delta_{\mathbf{D}}(\mathbf{x},\mathbf{u})\| \leq \epsilon_4. \quad (6.18)$$

Let $\mu_l(\theta(k))$ be the normalized fuzzy membership function defined
as (4.3). In the sequel, $\mu_l(\theta(k))$ is denoted as μ_l for convenience of
presentation. Then, the final output of the fuzzy system is inferred by
adopting a standard fuzzy inference method, i.e., using a center average
defuzzifier, singleton fuzzifier and product fuzzy inference

$$\begin{cases} \mathbf{x}(k+1) = \mathbf{A}(\mu)\mathbf{x}(k) + \mathbf{B}(\mu)\mathbf{u}(k) + \Delta_{\mathbf{A}}(\mathbf{x},\mathbf{u})\mathbf{x} + \Delta_{\mathbf{B}}(\mathbf{x},\mathbf{u})\mathbf{u} \\ \mathbf{y}(k) = \mathbf{C}(\mu)\mathbf{x}(k) + \mathbf{D}(\mu)\mathbf{u}(k) + \Delta_{\mathbf{C}}(\mathbf{x},\mathbf{u})\mathbf{x} + \Delta_{\mathbf{D}}(\mathbf{x},\mathbf{u})\mathbf{u} \end{cases}$$

$$(6.19)$$

where

$$\mathbf{A}(\mu) = \sum_{l=1}^{\kappa} \mu_l \mathbf{A}_l, \ \mathbf{B}(\mu) = \sum_{l=1}^{\kappa} \mu_l \mathbf{B}_l$$

$$\mathbf{C}(\mu) = \sum_{l=1}^{\kappa} \mu_l \mathbf{C}_l, \ \mathbf{D}(\mu) = \sum_{l=1}^{\kappa} \mu_l \mathbf{D}_l. \quad (6.20)$$

It is noted that the nonlinear process (6.1) can be approximated by the T-S fuzzy dynamic model (6.19). In what follows, the $\mathcal{L}_2$ fuzzy observer-based FD problem for systems (6.19) will be addressed based on piecewise-fuzzy Lypunov functions. To this end, the state-space partition is defined as $\{\mathcal{S}_i\}_{i\in\ell}$ and ℓ as the set of region indices. Similar to Chapter 5, the premise variable space $\mathcal{S} \in \mathcal{R}^p$ is partitioned into two kinds of regions: fuzzy (interpolation) regions and crisp (operating) regions. For each region $\mathcal{S}_i$, define

$$\Gamma(i) := \{l|\mu_l\left(\theta(k)\right) > 0,\ \theta(k) \in \mathcal{S}_i,\ i \in \ell\} \tag{6.21}$$

as the indices for the system matrices used in the interpolation within the region $\mathcal{S}_i$.

Now, nonlinear systems (6.1) can be equivalently expressed in the following piecewise-fuzzy form:

$$\begin{cases} \mathbf{x}(k+1) = (\mathcal{A}_i(\mu) + \Delta_\mathbf{A}(\mathbf{x}(k),\mathbf{u}(k)))\,\mathbf{x}(k) \\ \qquad\quad + (\mathcal{B}_i(\mu) + \Delta_\mathbf{B}(\mathbf{x}(k),\mathbf{u}(k)))\,\mathbf{u}(k) \\ \mathbf{y}(k) = (\mathcal{C}_i(\mu) + \Delta_\mathbf{C}(\mathbf{x}(k),\mathbf{u}(k)))\,\mathbf{x}(k) \\ \qquad\quad + (\mathcal{D}_i(\mu) + \Delta_\mathbf{D}(\mathbf{x}(k),\mathbf{u}(k)))\,\mathbf{u}(k) \\ \qquad\qquad\qquad \theta(k) \in \mathcal{S}_i,\ i \in \ell \end{cases} \tag{6.22}$$

where

$$\mathcal{A}_i(\mu) = \sum_{l\in\Gamma(i)} \mu_l \mathbf{A}_l,\quad \mathcal{B}_i(\mu) = \sum_{l\in\Gamma(i)} \mu_l \mathbf{B}_l$$

$$\mathcal{C}_i(\mu) = \sum_{l\in\Gamma(i)} \mu_l \mathbf{C}_l,\quad \mathcal{D}_i(\mu) = \sum_{l\in\Gamma(i)} \mu_l \mathbf{D}_l. \tag{6.23}$$

It is noted that the model of each local region is obtained by a blending of $l \in \Gamma(i)$ local models through fuzzy membership functions with $0 < \mu_l(\theta(k)) \leq 1$ and $\sum_{l\in\Gamma(i)} \mu_l(\theta(k)) = 1$.

Moreover, a new set Ω is defined to denote all possible region transitions

$$\Omega := \{(i,j)|\theta(k) \in \mathcal{S}_i,\ \theta(k+1) \in \mathcal{S}_j,\ i,j \in \ell\}. \tag{6.24}$$

Note that $j \neq i$ implies that the trajectories of premise variables $\theta(k)$ transit from the region $\mathcal{S}_i$ to $\mathcal{S}_j$. Otherwise, the trajectories of $\theta(k)$ stay in the same region $\mathcal{S}_i$.

6.2.2 Weighted Piecewise-Fuzzy Residual Generator

With the dynamic piecewise-fuzzy systems (6.22) on each region, the piecewise-fuzzy observer-based residual generator can be constructed as follows:

Region Rule i: IF $\theta(k) \in \mathcal{S}_i, i \in \ell$

Local Observer-based Residual Generator Rule $\mathfrak{R}^l$: IF $\theta_1(k)$ is N_1^l and $\theta_2(k)$ is N_2^l and $\cdots$ and $\theta_p(k)$ is N_p^l, THEN

$$\begin{cases} \hat{\mathbf{x}}(k+1) = \mathbf{A}_l\hat{\mathbf{x}}(k) + \mathbf{B}_l\mathbf{u}(t) + \mathbf{L}_{il}\left(\mathbf{y}(k) - \hat{\mathbf{y}}(k)\right) \\ \hat{\mathbf{y}}(k) = \mathbf{C}_l\hat{\mathbf{x}}(k) + \mathbf{D}_l\mathbf{u}(k) \\ \mathbf{r}(k) = \mathbf{y}(k) - \hat{\mathbf{y}}(k), \quad l \in \Gamma(i) \end{cases} \quad (6.25)$$

where $\hat{\mathbf{x}}(k) \in R^{k_x}$ denotes the state estimation; $\mathbf{r}(k) \in R^{k_y}$ represents the residual signal; $\mathbf{L}_{il}, l \in \Gamma(i), i \in \ell$ indicates the gain matrix of each local model in each local region, which will be determined later.

It turns out that the overall piecewise-fuzzy residual generator can be inferred in the following form:

$$\begin{cases} \hat{\mathbf{x}}(k+1) = \mathcal{A}_i(\mu)\hat{\mathbf{x}}(k) + \mathcal{B}_i(\mu)\mathbf{u}(k) + \mathcal{L}_i(\mu)\left(\mathbf{y}(k) - \hat{\mathbf{y}}(k)\right) \\ \hat{\mathbf{y}}(k) = \mathcal{C}_i(\mu)\hat{\mathbf{x}}(k) + \mathcal{D}_i(\mu)\mathbf{u}(k) \\ \mathbf{r}(k) = \mathbf{y}(k) - \hat{\mathbf{y}}(k), \quad \theta(k) \in \mathcal{S}_i, \ i \in \ell \end{cases} \quad (6.26)$$

where

$$\mathcal{L}_i(\mu) = \sum_{l \in \Gamma(i)} \mu_l \mathbf{L}_{il}, \ i \in \ell \quad (6.27)$$

is the overall gain matrix of each local region.

To realize a fault detection, an $\mathcal{L}_2$ observer-based residual generator will be studied first by addressing the existence condition given in Theorem 6.1 based on piecewise-fuzzy Lyapunov functions. For ease of presentation, we restrict our study in the sequel on system (6.1) satisfying the following $\mathcal{L}_2$ re-constructible condition

$$V\left(\mathbf{f}\left(\mathbf{x}, \mathbf{u}\right), \phi\left(\hat{\mathbf{x}}, \mathbf{u}, \mathbf{h}(\mathbf{x}, \mathbf{u})\right)\right) - V(\mathbf{x}, \hat{\mathbf{x}}) \leq -\mathbf{r}^T\mathbf{r} + \alpha^2\mathbf{u}^T\mathbf{u} \quad (6.28)$$

where $\alpha > 0$ is a given attenuation level.

Thus, provided the residual generator (6.26), it follows from Theorem 6.1 that an FD system (*FD system* 1) with evaluation window $[0, \tau]$ can be realized by defining the residual evaluation function

$$J(\mathbf{r}) = ||\mathbf{r}_\tau||_2^2 = \sum_{k=0}^{\tau} \mathbf{r}^T(k)\mathbf{r}(k) \qquad (6.29)$$

and setting the threshold

$$J_{\text{th}} = \alpha^2 ||\mathbf{u}_\tau||_2^2 + \sup_{\mathbf{x}(0),\hat{\mathbf{x}}(0)} V\left(\mathbf{x}(0), \hat{\mathbf{x}}(0)\right). \qquad (6.30)$$

Define a new residual signal

$$\tilde{\mathbf{r}}(k) := \mathbf{r}(k)/\alpha \qquad (6.31)$$

and consider the following design condition

$$\tilde{V}\left(f\left(\mathbf{x}, \mathbf{u}\right), \phi\left(\hat{\mathbf{x}}, \mathbf{u}, \mathbf{h}(\mathbf{x}, \mathbf{u})\right)\right) - \tilde{V}(\mathbf{x}, \hat{\mathbf{x}})$$
$$\leq -\tilde{\mathbf{r}}^T(k)\tilde{\mathbf{r}}(k) + \mathbf{u}^T(k)\mathbf{u}(k). \qquad (6.32)$$

It becomes evident that an alternative FD system (*FD system* 2), consists of residual generator (6.31), evaluation function $\tilde{J}(\tilde{\mathbf{r}}) = ||\tilde{\mathbf{r}}_\tau||_2^2$ and threshold

$$\tilde{J}_{\text{th}} = ||\mathbf{u}_\tau||_2^2 + \sup_{\mathbf{x}(0),\hat{\mathbf{x}}(0)} \tilde{V}\left(\mathbf{x}(0), \hat{\mathbf{x}}(0)\right) \qquad (6.33)$$

delivers the identical set of detectable faults like *FD system* 1

$$F_{(\mathbf{r}, J(\mathbf{r}), J_{\text{th}})} = F_{(\tilde{\mathbf{r}}, \tilde{J}(\tilde{\mathbf{r}}), \tilde{J}_{\text{th}})}. \qquad (6.34)$$

It is noted that the local dynamics of nonlinear systems can be significantly different. The existing difference among the local residual signals generated by residual generator (6.26) and (6.31) of each local region has not been fully taken into consideration. From the practical viewpoint, both *FD system* 1 and *FD system* 2 are working in a manner of worst case handling of uncertainties. The above conservative scheme may lead to a higher threshold and, as a result, (significant) reduction in fault detectability. This is the fact that motivates us to weight residual signals of each local regions differently, instead of by a constant, as described below

$$\begin{cases} \hat{\mathbf{x}}(k+1) = \mathcal{A}_i(\mu)\hat{\mathbf{x}}(k) + \mathcal{B}_i(\mu)\mathbf{u}(k) + \mathcal{L}_i(\mu)\left(\mathbf{y}(k) - \hat{\mathbf{y}}(k)\right) \\ \hat{\mathbf{y}}(k) = \mathcal{C}_i(\mu)\hat{\mathbf{x}}(k) + \mathcal{D}_i(\mu)\mathbf{u}(k) \\ \bar{\mathbf{r}}(k) = \omega_i\left(\mathbf{y}(k) - \hat{\mathbf{y}}(k)\right), \quad \theta(k) \in \mathcal{S}_i, \ i \in \ell \end{cases} \tag{6.35}$$

with $\omega_i > 0, i \in \ell$ denoting weighting constant of each local region.

Remark 6.1. *The essential idea of this approach is that the fuzzy logic will be used not only for residual generation but also for residual evaluation. In this sense, available knowledge of each local systems will be integrated into the residual evaluation and threshold computation.*

Remark 6.2. *It is noted that, the following weighted piecewise-fuzzy residual generator*

$$\bar{\mathbf{r}}(k) = \omega_i(\mu)\left(\mathbf{y}(k) - \hat{\mathbf{y}}(k)\right)$$
$$\omega_i(\mu) = \sum_{l \in \Gamma(i)} \mu_l \omega_{il}, \quad i \in \ell \tag{6.36}$$

can also be employed for FD purpose. However, enormous computation efforts are needed for the optimization of all the weighting factors $\omega_{il}, l \in \Gamma(i), i \in \ell$. From application perspective, we will focus on the weighted residual generator (6.35) in the sequel.

Consider the following design condition

$$\bar{V}\left(\mathbf{f}\left(\mathbf{x}, \mathbf{u}\right), \phi\left(\hat{\mathbf{x}}, \mathbf{u}, \mathbf{h}(\mathbf{x}, \mathbf{u})\right)\right) - \bar{V}(\mathbf{x}, \hat{\mathbf{x}})$$
$$\leq -\bar{\mathbf{r}}^T(k)\bar{\mathbf{r}}(k) + \mathbf{u}^T(k)\mathbf{u}(k) \tag{6.37}$$

by constructing residual generator (6.35), together with evaluation function $\bar{J}(\mathbf{r}) = ||\bar{\mathbf{r}}_\tau||_2^2$ and threshold

$$\bar{J}_{\text{th}} = ||\mathbf{u}_\tau||_2^2 + \sup_{\mathbf{x}(0), \hat{\mathbf{x}}(0)} \bar{V}\left(\mathbf{x}(0), \hat{\mathbf{x}}(0)\right) \tag{6.38}$$

an observer-based FD system (*FD system* 3) is realized.

Suppose that the weighting factors for residual generator (6.35) of *FD system 3* satisfy

$$\omega_i \geq \frac{1}{\alpha}, \quad i \in \ell \tag{6.39}$$

with α defined in (6.28). Then, it is evident that the fault detactability of *FD system* 3 will be enhanced compared with *FD system 1 and FD system 2*. That is to say

$$F_{(\mathbf{r},J(\mathbf{r}),J_{th})} = F_{(\tilde{\mathbf{r}},\tilde{J}(\tilde{\mathbf{r}}),\tilde{J}_{th})} \subseteq F_{(\bar{\mathbf{r}},\bar{J}(\bar{\mathbf{r}}),\bar{J}_{th})}. \tag{6.40}$$

To understand it, recall that the constant α represents the maximum influence of $\mathbf{u}(k)$ on $\mathbf{r}(k)$ and should cover the whole working range. *From this viewpoint, it is reasonable to suppose that there exists $\alpha_i = 1/\omega_i, i \in \ell$, which measures the maximum influence of $\mathbf{u}(k)$ on the residual signal of each local region such that $\alpha_i \leq \alpha$ or $\omega_i \geq 1/\alpha$. In this way, those local residual signals generated by residual generator (6.35) would be stronger weighted if the influence of $\mathbf{u}(k)$ on them is weaker. On the other hand, the local residual signal, on which the influence of $\mathbf{u}(k)$ is the strongest, is least weighted. This is the basic idea behind the piecewise-fuzzy observer-based FD scheme proposed in this study.*

6.2.3 An Integrated Design Scheme of FD Systems

In what follows, the determination of gain matrices $\mathbf{L}_{il}, l \in \Gamma(i), i \in \ell$ and weighting factors $\omega_i, i \in \ell$ for *FD system* 3 will be addressed by solving the design condition (6.37).

Define $\mathbf{e}(k) = \mathbf{x}(k) - \hat{\mathbf{x}}(k)$ and $\bar{\mathbf{x}}(k) = \begin{bmatrix} \mathbf{e}^T(k) & \mathbf{x}^T(k) \end{bmatrix}^T$, then the dynamics of the augmented system is governed by

$$\begin{cases} \bar{\mathbf{x}}(k+1) = (\bar{\mathcal{A}}_i(\mu) + \Delta_{\mathcal{A}_i}(\mu))\bar{\mathbf{x}}(k) + (\bar{\mathcal{B}}_i(\mu) + \Delta_{\mathcal{B}_i}(\mu))\mathbf{u}(k) \\ \bar{\mathbf{r}}(k) = \omega_i \left((\bar{\mathcal{C}}_i(\mu) + \Delta_{\mathcal{C}})\bar{\mathbf{x}}(k) + \Delta_{\mathcal{D}}\mathbf{u}(k) \right), \quad \theta(k) \in \mathcal{S}_i, \; i \in \ell \end{cases} \tag{6.41}$$

where

$$\bar{\mathcal{A}}_i(\mu) = \begin{bmatrix} \mathcal{A}_i(\mu) - \mathcal{L}_i(\mu)\mathcal{C}_i(\mu) & 0 \\ 0 & \mathcal{A}_i(\mu) \end{bmatrix}$$

$$\Delta_{\mathcal{A}_i}(\mu) = \begin{bmatrix} 0 & \Delta_{\mathbf{A}}(\mathbf{x},\mathbf{u}) - \mathcal{L}_i(\mu)\Delta_{\mathbf{C}}(\mathbf{x},\mathbf{u}) \\ 0 & \Delta_{\mathbf{A}}(\mathbf{x},\mathbf{u}) \end{bmatrix}$$

$$\bar{\mathcal{B}}_i(\mu) = \begin{bmatrix} 0 \\ \mathcal{B}_i(\mu) \end{bmatrix}, \quad \Delta_{\mathcal{B}_i}(\mu) = \begin{bmatrix} \Delta_{\mathbf{B}}(\mathbf{x},\mathbf{u}) - \mathcal{L}_i(\mu)\Delta_{\mathbf{D}}(\mathbf{x},\mathbf{u}) \\ \Delta_{\mathbf{B}}(\mathbf{x},\mathbf{u}) \end{bmatrix}$$

$$\bar{\mathcal{C}}_i(\mu) = \begin{bmatrix} \mathcal{C}_i(\mu) & 0 \end{bmatrix}, \quad \Delta_{\mathcal{C}} = \begin{bmatrix} 0 & \Delta_{\mathbf{C}}(\mathbf{x},\mathbf{u}) \end{bmatrix}, \quad \Delta_{\mathcal{D}} = \Delta_{\mathbf{D}}(\mathbf{x},\mathbf{u}).$$

The major design scheme for the observer-based residual generator (6.35) based on piecewise-fuzzy Lyapunov functions is then formulated in the following theorem [82].

Theorem 6.2. *Consider nonlinear systems (6.1), if there exist matrices*

$$\mathbf{P}_{il} > 0,\ \bar{\mathbf{L}}_{il},\ \mathbf{Y}_i = \begin{bmatrix} \mathbf{Y}_{i1} & \mathbf{Y}_{i2} \\ \mathbf{0} & \mathbf{Y}_{i3} \end{bmatrix},\ i \in \ell,\ l \in \Gamma(i) \qquad (6.42)$$

and constants $\xi_i > 0, \omega_i > 0, i \in \ell$, *such that the following LMIs are feasible*

$$\Xi_{ijmll} < \mathbf{0}, \quad i, j \in \ell \qquad (6.43)$$

$$\Xi_{ijmlp} + \Xi_{ijmpl} < \mathbf{0}, \quad p > l,\ i, j \in \ell \qquad (6.44)$$

where $m \in \Gamma(j),\ l, p \in \Gamma(i),\ (i,j) \in \Omega$ *and*

$$\Xi_{ijmlp} = \begin{bmatrix} -\omega_i^{-2}\mathbf{I} & \mathbf{0} & \bar{\mathbf{C}}_l & \mathbf{0} & \bar{\mathbf{N}} \\ \star & \mathbf{P}_{jm} - \mathbf{Y}_i - \mathbf{Y}_i^T & \bar{\mathbf{A}}_{ilp} & \bar{\mathbf{B}}_{il} & \bar{\mathbf{M}}_{il} \\ \star & \star & \mathbf{G}_i - \mathbf{P}_{il} & \mathbf{0} & \mathbf{0} \\ \star & \star & \star & -\mathbf{I} + \xi_i\lambda\mathbf{I} & \mathbf{0} \\ \star & \star & \star & \star & -\xi_i\mathbf{I} \end{bmatrix}$$

$$\mathbf{G}_i = \begin{bmatrix} \mathbf{0} & \mathbf{0} \\ \star & \xi_i\lambda\mathbf{I} \end{bmatrix},\ \bar{\mathbf{C}}_l = \begin{bmatrix} \mathbf{C}_l & \mathbf{0} \end{bmatrix},\ \bar{\mathbf{N}} = \begin{bmatrix} \mathbf{0} & \mathbf{I} \end{bmatrix}$$

$$\bar{\mathbf{A}}_{ilp} = \begin{bmatrix} \mathbf{Y}_{i1}\mathbf{A}_l - \bar{\mathbf{L}}_{il}\mathbf{C}_p & \mathbf{Y}_{i2}\mathbf{A}_l \\ \mathbf{0} & \mathbf{Y}_{i3}\mathbf{A}_l \end{bmatrix},\ \bar{\mathbf{B}}_{il} = \begin{bmatrix} \mathbf{Y}_{i2}\mathbf{B}_l \\ \mathbf{Y}_{i3}\mathbf{B}_l \end{bmatrix}$$

$$\bar{\mathbf{M}}_{il} = \begin{bmatrix} \mathbf{Y}_{i1} + \mathbf{Y}_{i2} & -\bar{\mathbf{L}}_{il} \\ \mathbf{Y}_{i3} & \mathbf{0} \end{bmatrix},\ \lambda = \epsilon_1^2 + \epsilon_2^2 + \epsilon_3^2 + \epsilon_4^2. \qquad (6.45)$$

Then, by employing $\mathbf{L}_{il} = \mathbf{Y}_{i1}^{-1}\bar{\mathbf{L}}_{il}$, *residual generator (6.35) results in*

$$\|\bar{\mathbf{r}}_\tau\|_2^2 < \|\mathbf{u}_\tau\|_2^2 + \bar{\mathbf{x}}^T(0)\bar{\mathcal{P}}_{i_0}\bar{\mathbf{x}}(0), \quad \theta(0) \in \mathcal{S}_{i_0},\ i_0 \in \ell \qquad (6.46)$$

with $\bar{\mathcal{P}}_{i_0} = \sum\limits_{l \in \Gamma(i_0)} \mu_l \mathbf{P}_{i_0 l}$.

Proof. Consider the following piecewise-fuzzy quadratic Lyapunov function

$$V(\bar{\mathbf{x}}(k)) = \bar{\mathbf{x}}^T(k)\mathcal{P}_i(\mu)\bar{\mathbf{x}}(k), \quad \theta(k) \in \mathcal{S}_i,\ i \in \ell \qquad (6.47)$$

where $\mathcal{P}_i(\mu) = \sum\limits_{l\in\Gamma(i)} \mu_l(\theta(k))\mathbf{P}_{il}, i \in \ell$ and $\mathbf{P}_{il} > 0, i \in \ell, l \in \Gamma(i)$ are symmetric positive definite matrices.

It follows directly from Theorem 6.1 that (6.46) is achieved if the system (6.1) is $\mathcal{L}_2$ re-constructable as (6.32). Suppose that (6.35) is the piecewise-fuzzy residual generator and (6.47) is the function $V(\cdot)$ in (6.32), thus it is equivalent to express the condition (6.32) into

$$V(\bar{\mathbf{x}}(k+1)) - V(\bar{\mathbf{x}}(k)) + \bar{\mathbf{r}}^T(k)\bar{\mathbf{r}}(k) - \mathbf{u}^T(k)\mathbf{u}(k) < \mathbf{0}. \qquad (6.48)$$

Assume $\theta(k+1) \in \mathcal{S}_j, (i,j) \in \Omega$, then by the substitution of (6.41) into (6.48), it is easy to see

$$\bar{\mathbf{x}}^T(k+1)\mathcal{P}_j(\mu^+)\bar{\mathbf{x}}(k+1) - \bar{\mathbf{x}}^T(k)\mathcal{P}_i(\mu)\bar{\mathbf{x}}(k) + \bar{\mathbf{r}}^T(k)\bar{\mathbf{r}}(k) - \mathbf{u}^T(k)\mathbf{u}(k)$$

$$= \begin{bmatrix} \bar{\mathbf{x}}(k) \\ \mathbf{u}(k) \end{bmatrix}^T \left(\begin{bmatrix} \bar{A}_i^T(\mu) + \Delta_{\mathcal{A}_i}^T(\mu) \\ \bar{B}_i^T(\mu) + \Delta_{\mathcal{B}_i}^T(\mu) \end{bmatrix} \mathcal{P}_j(\mu^+)(\star) \right.$$

$$\left. + \omega_i^2 \begin{bmatrix} \bar{C}_i^T(\mu) + \Delta_{\mathcal{C}}^T \\ \Delta_{\mathcal{D}}^T \end{bmatrix} (\star) - \begin{bmatrix} \mathcal{P}_i(\mu) & \mathbf{0} \\ \mathbf{0} & \mathbf{I} \end{bmatrix} \right) \begin{bmatrix} \bar{\mathbf{x}}(k) \\ \mathbf{u}(k) \end{bmatrix}. \qquad (6.49)$$

Here, $\mathcal{P}_j(\mu^+) = \sum\limits_{m\in\Gamma(j)} \mu_m^+(\theta(k+1))\mathbf{P}_{jm}, j \in \ell$. By applying Schur Complement to (6.49), it is easy to see that the following inequality implies (6.48)

$$\begin{bmatrix} -\omega_i^{-2}\mathbf{I} & \mathbf{0} & \bar{C}_i(\mu) + \Delta_\mathcal{C} & \Delta_\mathcal{D} \\ \star & -(\mathcal{P}_j(\mu^+))^{-1} & \bar{A}_i(\mu) + \Delta_{\mathcal{A}_i}(\mu) & \bar{B}_i(\mu) + \Delta_{\mathcal{B}_i}(\mu) \\ \star & \star & -\mathcal{P}_i(\mu) & \mathbf{0} \\ \star & \star & \star & -\mathbf{I} \end{bmatrix} < \mathbf{0}$$

$$i, j \in \ell, \ (i,j) \in \Omega. \quad (6.50)$$

Note that the inverse of the piecewise-fuzzy Lyapunov matrices $(\mathcal{P}_j(\mu^+))^{-1}$ is involved in (6.50). To eliminate the coupling between the Lyapunov matrices $\mathcal{P}_j(\mu^+)$ and system matrices, we perform a congruence transformation to (6.50) by $\text{diag}(\mathbf{I}, \mathbf{Y}_i, \mathbf{I}, \mathbf{I})$ with $Y_i, i \in \ell$ as a set of nonsingular matrices. It follows from

$$(\mathcal{P}_j(\mu^+) - \mathbf{Y}_i)(\mathcal{P}_j(\mu^+))^{-1}(\mathcal{P}_j(\mu^+) - \mathbf{Y}_i)^T$$

$$= \mathbf{Y}_i(\mathcal{P}_j(\mu^+))^{-1}\mathbf{Y}_i^T + \mathcal{P}_j(\mu^+) - \mathbf{Y}_i - \mathbf{Y}_i^T \geq 0 \qquad (6.51)$$

that

$$-\mathbf{Y}_i\left(\mathcal{P}_j(\mu^+)\right)^{-1}\mathbf{Y}_i^T \le \mathcal{P}_j(\mu^+) - \mathbf{Y}_i - \mathbf{Y}_i^T. \qquad (6.52)$$

Therefore, based on (6.52), the following inequality leads to (6.50)

$$\begin{bmatrix} -\omega_i^{-2}\mathbf{I} & \mathbf{0} & \bar{\mathcal{C}}_i(\mu) + \Delta_{\mathcal{C}} \\ \star & \mathcal{P}_j(\mu^+) - \mathbf{Y}_i - \mathbf{Y}_i^T & \mathbf{Y}_i\left(\bar{\mathcal{A}}_i(\mu) + \Delta_{\mathcal{A}_i}(\mu)\right) \\ \star & \star & -\mathcal{P}_i(\mu) \\ \star & \star & \star \end{bmatrix}$$

$$\begin{matrix} \Delta_{\mathcal{D}} \\ \mathbf{Y}_i\left(\bar{\mathcal{B}}_i(\mu) + \Delta_{\mathcal{B}_i}(\mu)\right) \\ \mathbf{0} \\ -\mathbf{I} \end{matrix} \Bigg] < \mathbf{0}, \ \ i,j \in \ell, \ (i,j) \in \Omega. \quad (6.53)$$

It can be easily observed from (6.53) that the non-singularity of $\mathbf{Y}_i, i \in \ell$ is ensured by $\mathcal{P}_j(\mu^+) - \mathbf{Y}_i - \mathbf{Y}_i^T < 0$.

Now, by expanding the fuzzy-basis functions, the left-hand-side (LHS) of (6.53) can be expressed as follows:

$$\text{LHS}(6.53) = \sum_{m\in\Gamma(j)} \sum_{l\in\Gamma(i)} \sum_{p\in\Gamma(i)} \mu_m^+\mu_l\mu_p \mathbf{\Psi}_{ijmlp}$$

$$= \sum_{m\in\Gamma(j)} \sum_{l\in\Gamma(i)} \sum_{p>l, p\in\Gamma(i)} \mu_m^+\mu_l\mu_p\left(\mathbf{\Psi}_{ijmlp} + \mathbf{\Psi}_{ijmpl}\right)$$

$$+ \sum_{m\in\Gamma(j)} \sum_{l\in\Gamma(i)} \mu_m^+\mu_l^2 \mathbf{\Psi}_{ijmll}, \ \ i,j \in \ell, \ (i,j) \in \Omega \quad (6.54)$$

where

$$\mathbf{\Psi}_{ijmlp} = \begin{bmatrix} -\omega_i^{-2}\mathbf{I} & \mathbf{0} & \bar{\mathbf{C}}_l + \Delta_{\mathcal{C}} \\ \star & \mathbf{P}_{jm} - \mathbf{Y}_i - \mathbf{Y}_i^T & \mathbf{Y}_i(\tilde{\mathbf{A}}_{ilp} + \tilde{\Delta}_{\mathbf{A}_{il}}) \\ \star & \star & -\mathbf{P}_{il} \\ \star & \star & \star \end{bmatrix}$$

$$\begin{matrix} \Delta_{\mathcal{D}} \\ \mathbf{Y}_i(\tilde{\mathbf{B}}_l + \tilde{\Delta}_{\mathbf{B}_{il}}) \\ \mathbf{0} \\ -\mathbf{I} \end{matrix} \Bigg]$$

$$\tilde{\mathbf{A}}_{ilp} = \begin{bmatrix} \mathbf{A}_l - \mathbf{L}_{il}\mathbf{C}_p & \mathbf{0} \\ \mathbf{0} & \mathbf{A}_l \end{bmatrix}, \ \tilde{\Delta}_{\mathbf{A}_{il}} = \begin{bmatrix} \mathbf{0} & \Delta_{\mathbf{A}}(x,u) - \mathbf{L}_{il}\Delta_{\mathbf{C}}(x,u) \\ \mathbf{0} & \Delta_{\mathbf{A}}(x,u) \end{bmatrix}$$

$$\tilde{\mathbf{B}}_l = \begin{bmatrix} \mathbf{0} \\ \mathbf{B}_l \end{bmatrix}, \ \tilde{\Delta}_{\mathbf{B}_{il}} = \begin{bmatrix} \Delta_{\mathbf{B}}(\mathbf{x}, \mathbf{u}) - \mathbf{L}_{il}\Delta_{\mathbf{D}}(\mathbf{x}, \mathbf{u}) \\ \Delta_{\mathbf{B}}(\mathbf{x}, \mathbf{u}) \end{bmatrix}, \ \bar{\mathbf{C}}_l = \begin{bmatrix} \mathbf{C}_l & \mathbf{0} \end{bmatrix}.$$

Then, it is easy to see that the following inequalities imply (6.53)

$$\Psi_{ijmll} < \mathbf{0}, \quad i, j \in \ell, \ (i, j) \in \Omega \tag{6.55}$$

$$\Psi_{ijmlp} + \Psi_{ijmpl} < \mathbf{0}, \quad p > l, \ i, j \in \ell, \ (i, j) \in \Omega \tag{6.56}$$

where $m \in \Gamma(j), l, p \in \Gamma(i)$.

For simplicity, we just consider the proof of the most complex case (6.56). By extracting the parameter uncertainties, one can show that

$$\Psi_{ijmlp} + \Psi_{ijmpl}$$
$$= \Sigma_{ijmlp} + \Sigma_{ijmpl} + \mathbf{H}_{il}\bar{\Delta} + \bar{\Delta}^T\mathbf{H}_{il}^T + \mathbf{H}_{ip}\bar{\Delta} + \bar{\Delta}^T\mathbf{H}_{ip}^T < \mathbf{0}$$
$$p > l, \ m \in \Gamma(j), \ l, p \in \Gamma(i), \ i, j \in \ell, \ (i, j) \in \Omega \tag{6.57}$$

where

$$\Sigma_{ijmlp} = \begin{bmatrix} -\omega_i^{-2}\mathbf{I} & \mathbf{0} & \tilde{\mathbf{C}}_l & \mathbf{0} \\ \star & P_{jm} - \mathbf{Y}_i - \mathbf{Y}_i^T & \mathbf{Y}_i\tilde{\mathbf{A}}_{ilp} & \mathbf{Y}_i\tilde{\mathbf{B}}_l \\ \star & \star & -\mathbf{P}_{il} & \mathbf{0} \\ \star & \star & \star & -\mathbf{I} \end{bmatrix}$$

$$\mathbf{H}_{il} = \begin{bmatrix} \bar{\mathbf{N}} \\ \mathbf{Y}_i\mathbf{M}_{il} \\ \mathbf{0} \\ \mathbf{0} \end{bmatrix}, \mathbf{M}_{il} = \begin{bmatrix} \mathbf{I} & -\mathbf{L}_{il} \\ \mathbf{I} & \mathbf{0} \end{bmatrix}, \bar{\mathbf{N}} = \begin{bmatrix} \mathbf{0} & \mathbf{I} \end{bmatrix}$$

$$\bar{\Delta} = \begin{bmatrix} \mathbf{0} & \mathbf{0} \begin{bmatrix} \mathbf{0} & \Delta_{\mathbf{A}}(\mathbf{x}, \mathbf{u}) & \Delta_{\mathbf{B}}(\mathbf{x}, \mathbf{u}) \\ \mathbf{0} & \mathbf{0} \begin{bmatrix} \mathbf{0} & \Delta_{\mathbf{C}}(\mathbf{x}, \mathbf{u}) & \Delta_{\mathbf{D}}(\mathbf{x}, \mathbf{u}) \end{bmatrix} \end{bmatrix}. \tag{6.58}$$

It is known that, for any $\xi_i > 0, i \in \ell$, one obtains

$$(\mathbf{H}_{il} + \mathbf{H}_{ip})^T \bar{\Delta} + \bar{\Delta}^T (\mathbf{H}_{il} + \mathbf{H}_{ip})$$
$$\leq \frac{1}{2\xi_i} (\mathbf{H}_{il} + \mathbf{H}_{ip}) (\mathbf{H}_{il} + \mathbf{H}_{ip})^T + 2\xi_i\bar{\Delta}^T\bar{\Delta}. \tag{6.59}$$

Notice that for $\lambda = \epsilon_1^2 + \epsilon_2^2 + \epsilon_3^2 + \epsilon_4^2$, it holds

$$\begin{bmatrix} \Delta_{\mathbf{A}}(\mathbf{x}, \mathbf{u}) & \Delta_{\mathbf{B}}(\mathbf{x}, \mathbf{u}) \\ \Delta_{\mathbf{C}}(\mathbf{x}, \mathbf{u}) & \Delta_{\mathbf{D}}(\mathbf{x}, \mathbf{u}) \end{bmatrix}^T (\star) \leq \lambda\mathbf{I}. \tag{6.60}$$

Then it follows directly from (6.59), (6.60) and Schur Complement that (6.57) is achieved if the following inequality is feasible

$$\left[\begin{array}{cc} \Sigma_{ijmlp} + \Sigma_{ijmpl} + 2\mathbf{W}_i & \mathbf{H}_{il} + \mathbf{H}_{ip} \\ \star & -2\xi_i\mathbf{I} \end{array} \right] < \mathbf{0}$$

$$p > l, \ m \in \Gamma(j), \ l,p \in \Gamma(i), \ i,j \in \ell, \ (i,j) \in \Omega \qquad (6.61)$$

where

$$\mathbf{W}_i = \left[\begin{array}{cccc} \mathbf{0} & \mathbf{0} & \mathbf{0} & \mathbf{0} \\ \star & \mathbf{0} & \mathbf{0} & \mathbf{0} \\ \star & \star & \mathbf{G}_i & \mathbf{0} \\ \star & \star & \star & \xi_i\lambda\mathbf{I} \end{array} \right], \ \mathbf{G}_i = \left[\begin{array}{cc} \mathbf{0} & \mathbf{0} \\ \star & \xi_i\lambda\mathbf{I} \end{array} \right]. \qquad (6.62)$$

In addition, it is noted that (6.61) can be equivalenty re-expressed as

$$\mathbf{\Phi}_{ijmlp} + \mathbf{\Phi}_{ijmpl} < \mathbf{0}, \quad p > l, \ m \in \Gamma(j), \ l,p \in \Gamma(i), \ i,j \in \ell, \ (i,j) \in \Omega$$
$$(6.63)$$

where

$$\mathbf{\Phi}_{ijmlp} = \left[\begin{array}{ccccc} -\omega_i^{-2}\mathbf{I} & \mathbf{0} & \tilde{\mathbf{C}}_l & \mathbf{0} & \bar{\mathbf{N}} \\ \star & \mathbf{P}_{jm} - \mathbf{Y}_i - \mathbf{Y}_i^T & \mathbf{Y}_i\tilde{\mathbf{A}}_{ilp} & \mathbf{Y}_i\tilde{\mathbf{B}}_l & \mathbf{Y}_i\mathbf{M}_{il} \\ \star & \star & \mathbf{G}_i - \mathbf{P}_{il} & \mathbf{0} & \mathbf{0} \\ \star & \star & \star & -\mathbf{I} + \xi_i\lambda\mathbf{I} & \mathbf{0} \\ \star & \star & \star & \star & -\xi_i\mathbf{I} \end{array} \right]$$
$$(6.64)$$

Moreover, defining the matrices $\mathbf{Y}_i$ as in (6.42) and $\bar{\mathbf{L}}_{il} = \mathbf{Y}_{i1}\mathbf{L}_{il}$, together with the inequality (6.44), it becomes evident that (6.56) holds. On the other hand, it can be proved that the inequality (6.43) implies (6.55). Thus, the proof is completed. $\qquad\square$

Remark 6.3. *In practice, we often meet the case where the output behavior of the systems is globally linear, i.e.,*

$$\mathbf{C}_l = \mathbf{C}, \ \mathbf{D}_l = \mathbf{D}, \quad l = 1, \cdots, \kappa. \qquad (6.65)$$

In this case, the gain matrices and weighting factors of residual generator (6.35) can be attained by solving the set of LMIs in (6.43).

Algorithm 3 Solution of "gain-schedule" table

1: Find max $\omega_i = \omega_i^*$, $i \in \ell$ by solving (6.43)-(6.44) iteratively
2: Set $\omega_i = \omega_i^*$, $i \in \ell$

We summarize the solution of "gain-schedule" table in Algorithm 3. Now we are able to apply the following on-line implementation of the piecewise-fuzzy observer-based FD systems for nonlinear processes (6.1):

- Run the residual generator (6.35)

- Set

$$\bar{J}_{\text{th}} = \|\mathbf{u}_\tau\|_2^2 + \max_{\mathbf{x}(0),\hat{\mathbf{x}}(0)} \bar{\mathbf{x}}^T(0)\bar{\mathcal{P}}_{i_0}\bar{\mathbf{x}}(0) \qquad (6.66)$$

- Set the decision logic

$$\begin{cases} \bar{J} = \|\bar{\mathbf{r}}_\tau\|_2^2 > \bar{J}_{\text{th}} \Longrightarrow \text{faulty} \\ \bar{J} = \|\bar{\mathbf{r}}_\tau\|_2^2 \leq \bar{J}_{\text{th}} \Longrightarrow \text{fault-free.} \end{cases} \qquad (6.67)$$

Remark 6.4. *It is worth pointing out that by defining $\mathbf{P}_{il} \equiv \mathbf{P} > 0$, $l \in \Gamma(i)$, $i \in \ell$, the determination of the gain matrices and weighting factors for residual generator (6.35) can be realized via the common Lyapunov function-based approach. Nevertheless, the computation complexity for solving the set of LMIs (6.43)-(6.44) is reduced. However, in this way, the design conservatism will increase and the resulting FD performance shall become worse. In Section 6.4, an example is given to show the comparison between the FD performance of common Lyapunov function-based approach and the proposed piecewise-fuzzy Lyapunov function-based approach.*

Remark 6.5. *We would like to mention that our study is also motivated by the observations and experiences in some industrial projects. It is noticed that although the limit monitoring and trend analysis techniques are widely used in practice for the FD purpose, the thresholds (limits) are usually determined by tests, simulation or even based on experiences. Moreover, to deal with nonlinearity, establishing a local region dependent "gain-schedule" table is a typical way adopted in industrial processes.*

6.3 $\mathcal{L}_2$ Robust Piecewise-Fuzzy Observer-based FD Systems

In this section, we will address the robust piecewise-fuzzy observer-based FD issues for nonlinear systems with external disturbances, as described by

$$\begin{cases} \mathbf{x}(k+1) = \mathbf{f}(\mathbf{x}(k), \mathbf{u}(k)) + \mathbf{g}(\mathbf{x}(k), \mathbf{u}(k))\mathbf{d}(k) \\ \mathbf{y}(k) = \mathbf{h}(\mathbf{x}(k), \mathbf{u}(k)) + \mathbf{v}(\mathbf{x}(k), \mathbf{u}(k))\mathbf{d}(k) \end{cases} \tag{6.68}$$

where $\mathbf{g}(\mathbf{x}(k), \mathbf{u}(k))$ and $\mathbf{v}(\mathbf{x}(k), \mathbf{u}(k))$ are continuously differentiable nonlinear matrix functions with appropriate dimensions. $\mathbf{d}(k) \in \mathcal{R}^{n_d}$ denotes the disturbances which is $\mathcal{L}_2$-bounded with

$$\|\mathbf{d}_\tau\|_2 \leq \delta_\mathbf{d}. \tag{6.69}$$

Before proceeding further, we first extend the existence condition of $\mathcal{L}_2$ robust observer-based FD systems to nonlinear processes (6.68).

Definition 6.2. *System (6.68) is weakly output re-constructible if there exist (i) functions $\phi : \mathcal{R}^{n_x} \times \mathcal{R}^{n_u} \times \mathcal{R}^{n_y} \to \mathcal{R}^{n_x}$ and $\psi : \mathcal{R}^{n_y} \to \mathcal{R}^{n_r}$; (ii) functions $V : \mathcal{R}^{n_x} \times \mathcal{R}^{n_x} \to \mathcal{R}^+, \varphi_1(\cdot) \in \mathcal{K}, \varphi_2(\cdot) \in \mathcal{K}_\infty, \varphi_3(\cdot) \in \mathcal{K}_\infty$ and positive constant δ such that $\forall \mathbf{x}(0), \hat{\mathbf{x}}(0) \in \mathcal{B}_\delta$*

$$\begin{aligned} &V\left(\mathbf{f}\left(\mathbf{x}, \mathbf{u}\right), \phi\left(\hat{\mathbf{x}}, \mathbf{u}, \mathbf{h}(\mathbf{x}, \mathbf{u})\right)\right) - V(\mathbf{x}, \hat{\mathbf{x}}) \\ &\leq -\varphi_1(\|\mathbf{r}\|) + \varphi_2(\|\mathbf{u}\|) + \varphi_3(\|\mathbf{d}\|) \\ &\mathbf{r} = \psi(\mathbf{y}(k) - \mathbf{h}(\hat{\mathbf{x}}(k), \mathbf{u}(k))) \end{aligned} \tag{6.70}$$

Remark 6.6. *Along with the similar line of the proof of Theorem 6.1, if (6.70) holds, we have*

$$\sum_{k=0}^{\tau} \varphi_1(\|\mathbf{r}(k)\|) \leq \sum_{k=0}^{\tau} \varphi_2(\|\mathbf{u}(k)\|) + \sum_{k=0}^{\tau} \varphi_3(\|\mathbf{d}(k)\|) + V(\mathbf{x}(0), \hat{\mathbf{x}}(0)). \tag{6.71}$$

Thus, an $\mathcal{L}_2$ robust observer-based FD system for nonlinear processes (6.68) can be realized.

As discussed in Chapter 4, for any given positive constant ϵ_1, ϵ_2, ϵ_3, ϵ_4, ϵ_5, ϵ_6, one can find a fuzzy law such that nonlinear systems (6.68) can be approximated by the following T-S fuzzy models with some norm-bounded uncertainties:

Plant rule $\Re^l$: *IF* $\theta_1(k)$ *is* N_1^l *and* $\theta_2(k)$ *is* N_2^l *and* $\cdots$ *and* $\theta_p(k)$ *is* N_p^l, *THEN*

$$
\begin{cases}
\mathbf{x}(k+1) = \mathbf{A}_l\mathbf{x}(k) + \mathbf{B}_l\mathbf{u}(k) + \mathbf{E}_l\mathbf{d}(k) + \Delta_{\mathbf{A}}(\mathbf{x}(k),\mathbf{u}(k))\mathbf{x}(k) \\
\qquad + \Delta_{\mathbf{B}}(\mathbf{x}(k),\mathbf{u}(k))\mathbf{u}(k) + \Delta_{\mathbf{E}}(\mathbf{x}(k),\mathbf{u}(k))\mathbf{d}(k) \\
\mathbf{y}(k) = \mathbf{C}_l\mathbf{x}(k) + \mathbf{D}_l\mathbf{u}(k) + \mathbf{F}_l\mathbf{d}(k) + \Delta_{\mathbf{C}}(\mathbf{x}(k),\mathbf{u}(k))\mathbf{x}(k) \\
\qquad + \Delta_{\mathbf{D}}(\mathbf{x}(k),\mathbf{u}(k))\mathbf{u}(k) + \Delta_{\mathbf{F}}(\mathbf{x}(k),\mathbf{u}(k))\mathbf{d}(k) \\
\qquad\qquad\qquad\qquad\qquad l \in \{1,2,\cdots,\kappa\}
\end{cases}
$$

where $\mathbf{A}_l, \mathbf{B}_l, \mathbf{C}_l, \mathbf{D}_l, \mathbf{E}_l$ and $\mathbf{F}_l$ are matrices with appropriate dimensions and

$$
\begin{aligned}
\|\Delta_{\mathbf{A}}(\mathbf{x},\mathbf{u})\| \leq \epsilon_1, \quad \|\Delta_{\mathbf{B}}(\mathbf{x},\mathbf{u})\| \leq \epsilon_2, \quad \|\Delta_{\mathbf{E}}(\mathbf{x},\mathbf{u})\| \leq \epsilon_5 \\
\|\Delta_{\mathbf{C}}(\mathbf{x},\mathbf{u})\| \leq \epsilon_3, \quad \|\Delta_{\mathbf{D}}(\mathbf{x},\mathbf{u})\| \leq \epsilon_4, \quad \|\Delta_{\mathbf{F}}(\mathbf{x},\mathbf{u})\| \leq \epsilon_6.
\end{aligned} \tag{6.72}
$$

Thus the fuzzy system can be inferred as follows:

$$
\begin{cases}
\mathbf{x}(k+1) = \sum_{l=1}^{\kappa} \mu_l(\theta(k))\,(\mathbf{A}_l\mathbf{x}(k) + \mathbf{B}_l\mathbf{u}(k) + \mathbf{E}_l\mathbf{d}(k)) \\
\qquad + \Delta_{\mathbf{A}}(\mathbf{x}(k),\mathbf{u}(k))\mathbf{x}(k) + \Delta_{\mathbf{B}}(\mathbf{x}(k),\mathbf{u}(k))\mathbf{u}(k) \\
\qquad + \Delta_{\mathbf{E}}(\mathbf{x}(k),\mathbf{u}(k))\mathbf{d}(k) \\
\mathbf{y}(k) = \sum_{l=1}^{\kappa} \mu_l(\theta(k))\,(\mathbf{C}_l\mathbf{x}(k) + \mathbf{D}_l\mathbf{u}(k) + \mathbf{F}_l\mathbf{d}(k)) \\
\qquad + \Delta_{\mathbf{C}}(\mathbf{x}(k),\mathbf{u}(k))\mathbf{x}(k) + \Delta_{\mathbf{D}}(\mathbf{x}(k),\mathbf{u}(k))\mathbf{u}(k) \\
\qquad + \Delta_{\mathbf{F}}(\mathbf{x}(k),\mathbf{u}(k))\mathbf{d}(k).
\end{cases}
$$

Based on the same state-space partition approach in the previous section, nonlinear systems (6.68) can be equivalently re-written in the following piecewise-fuzzy form:

$$
\begin{cases}
\mathbf{x}(k+1) = \mathcal{A}_i(\mu)\mathbf{x}(k) + \mathcal{B}_i(\mu)\mathbf{u}(k) + \mathcal{E}_i(\mu)\mathbf{d}(k) + \Delta_{\mathbf{A}}(\mathbf{x}(k),\mathbf{u}(k))\mathbf{x}(k) \\
\qquad + \Delta_{\mathbf{B}}(\mathbf{x}(k),\mathbf{u}(k))\mathbf{u}(k) + \Delta_{\mathbf{E}}(\mathbf{x}(k),\mathbf{u}(k))\mathbf{d}(k) \\
\mathbf{y}(k) = \mathcal{C}_i(\mu)\mathbf{x}(k) + \mathcal{D}_i(\mu)\mathbf{u}(k) + \mathcal{F}_i(\mu)\mathbf{d}(k) + \Delta_{\mathbf{C}}(\mathbf{x}(k),\mathbf{u}(k))\mathbf{x}(k) \\
\qquad + \Delta_{\mathbf{D}}(\mathbf{x}(k),\mathbf{u}(k))\mathbf{u}(k) + \Delta_{\mathbf{F}}(\mathbf{x}(k),\mathbf{u}(k))\mathbf{d}(k) \\
\qquad\qquad\qquad\qquad\qquad \theta(k) \in \mathbf{S}_i,\ i \in \ell
\end{cases}
$$

$$\tag{6.73}$$

with

$$\mathcal{E}_i(\mu) = \sum_{l \in \Gamma(i)} \mu_l \mathbf{E}_l, \quad \mathcal{F}_i(\mu) = \sum_{l \in \Gamma(i)} \mu_l \mathbf{F}_l. \tag{6.74}$$

By employing the same weighted piecewise-fuzzy observer-based residual generator as in (6.35), the overall dynamics of residual generator can be further represented as follows:

$$\begin{cases} \bar{\mathbf{x}}(k+1) = (\bar{\mathcal{A}}_i(\mu) + \Delta_{\mathcal{A}_i}(\mu))\bar{\mathbf{x}}(k) + (\bar{\mathcal{B}}_i(\mu) + \Delta_{\mathcal{B}_i}(\mu))\mathbf{z}(k) \\ \bar{\mathbf{r}}(k) = \omega_i \left((\bar{\mathcal{C}}_i(\mu) + \Delta_{\mathcal{C}})\bar{\mathbf{x}}(k) + \left(\bar{\mathcal{D}}_i(\mu) + \Delta_{\mathcal{D}} \right) \mathbf{z}(k) \right) \end{cases} \tag{6.75}$$

where

$$\mathbf{e}(k) = \mathbf{x}(k) - \hat{\mathbf{x}}(k), \ \bar{\mathbf{x}}(k) = \begin{bmatrix} \mathbf{e}^T(k) & \mathbf{x}^T(k) \end{bmatrix}^T, \ \mathbf{z}(k) = \begin{bmatrix} \mathbf{u}^T(k) & \mathbf{d}^T(k) \end{bmatrix}^T$$

$$\bar{\mathcal{A}}_i(\mu) = \begin{bmatrix} \mathcal{A}_i(\mu) - \mathcal{L}_i(\mu)\mathcal{C}_i(\mu) & 0 \\ 0 & \mathcal{A}_i(\mu) \end{bmatrix}$$

$$\bar{\mathcal{B}}_i(\mu) = \begin{bmatrix} 0 & \mathcal{E}_i(\mu) - \mathcal{L}_i(\mu)\mathcal{F}_i(\mu) \\ \mathcal{B}_i(\mu) & \mathcal{E}_i(\mu) \end{bmatrix}$$

$$\Delta_{\mathcal{A}_i}(\mu) = \begin{bmatrix} 0 & \Delta_{\mathbf{A}}(\mathbf{x}, \mathbf{u}) - \mathcal{L}_i(\mu)\Delta_{\mathbf{C}}(\mathbf{x}, \mathbf{u}) \\ 0 & \Delta_{\mathbf{A}}(\mathbf{x}, \mathbf{u}) \end{bmatrix}$$

$$\Delta_{\mathcal{B}_i}(\mu) = \begin{bmatrix} \Delta_{\mathbf{B}}(\mathbf{x}, \mathbf{u}) - \mathcal{L}_i(\mu)\Delta_{\mathbf{D}}(\mathbf{x}, \mathbf{u}) \\ \Delta_{\mathbf{B}}(\mathbf{x}, \mathbf{u}) \end{bmatrix}$$

$$\bar{\mathcal{C}}_i(\mu) = \begin{bmatrix} \mathcal{C}_i(\mu) & 0 \end{bmatrix}, \ \bar{\mathcal{D}}_i(\mu) = \begin{bmatrix} 0 & \mathcal{F}_i(\mu) \end{bmatrix}$$

$$\Delta_{\mathcal{C}} = \begin{bmatrix} 0 & \Delta_{\mathbf{C}}(\mathbf{x}, \mathbf{u}) \end{bmatrix}, \ \Delta_{\mathcal{D}} = \begin{bmatrix} \Delta_{\mathbf{D}}(\mathbf{x}, \mathbf{u}) & \Delta_{\mathbf{F}}(\mathbf{x}, \mathbf{u}) \end{bmatrix}. \tag{6.76}$$

Then, the following results for robust residual generation is presented.

Theorem 6.3. *Consider nonlinear systems (6.68) and residual generator in the form of (6.35). It holds that*

$$\|\bar{\mathbf{r}}\|_2^2 < \|\mathbf{u}_\tau\|_2^2 + \|\mathbf{d}_\tau\|_2^2 + \bar{\mathbf{x}}^T(0)\bar{\mathbf{P}}_{i_0}\bar{\mathbf{x}}(0), \quad \theta(0) \in \mathbf{S}_{i_0}, \ i_0 \in \ell \tag{6.77}$$

with $\bar{\mathcal{P}}_{i_0} = \sum_{l \in \Gamma(i_0)} \mu_l \mathbf{P}_{i_0 l}$, *if there exist matrices*

$$\mathbf{P}_{il} > 0, \ \bar{\mathbf{L}}_{il}, \ \mathbf{Y}_i = \begin{bmatrix} \mathbf{Y}_{i1} & \mathbf{Y}_{i2} \\ 0 & \mathbf{Y}_{i3} \end{bmatrix}, \ i \in \ell, \ l \in \Gamma(i) \tag{6.78}$$

and constants $\xi_i > 0, \omega_i > 0, i \in \ell$, such that the following LMIs are feasible

$$\Theta_{ijmll} < \mathbf{0}, \quad i, j \in \ell \tag{6.79}$$

$$\Theta_{ijmlp} + \Theta_{ijmpl} < \mathbf{0}, \quad p > l, \ i, j \in \ell \tag{6.80}$$

where $m \in \Gamma(j), l, p \in \Gamma(i), (i, j) \in \Omega$ and

$$\Theta_{ijmlp} = \begin{bmatrix} -\omega_i^{-2}\mathbf{I} & \mathbf{0} & \bar{\mathbf{C}}_l & \bar{\mathbf{D}}_l & \bar{\mathbf{N}} \\ \star & \mathbf{P}_{jm} - \mathbf{Y}_i - \mathbf{Y}_i^T & \bar{\mathbf{A}}_{ilp} & \bar{\mathbf{B}}_{il} & \bar{\mathbf{M}}_{il} \\ \star & \star & \mathbf{G}_i - \mathbf{P}_{il} & \mathbf{0} & \mathbf{0} \\ \star & \star & \star & -\mathbf{I} + \xi_i \lambda \mathbf{I} & \mathbf{0} \\ \star & \star & \star & \star & -\xi_i \mathbf{I} \end{bmatrix}$$

$$\bar{\mathbf{A}}_{ilp} = \begin{bmatrix} \mathbf{Y}_{i1}\mathbf{A}_l - \bar{\mathbf{L}}_{il}\mathbf{C}_p & \mathbf{Y}_{i2}\mathbf{A}_l \\ \mathbf{0} & \mathbf{Y}_{i3}\mathbf{A}_l \end{bmatrix}, \quad \bar{\mathbf{C}}_l = \begin{bmatrix} \mathbf{C}_l & \mathbf{0} \end{bmatrix}$$

$$\bar{\mathbf{B}}_{il} = \begin{bmatrix} \mathbf{Y}_{i2}\mathbf{B}_l & \mathbf{Y}_{i1}\mathbf{E}_l + \mathbf{Y}_{i2}\mathbf{E}_l - \bar{\mathbf{L}}_{il}\mathbf{F}_l \\ \mathbf{Y}_{i3}\mathbf{B}_l & \mathbf{Y}_{i3}\mathbf{E}_l \end{bmatrix}$$

$$\bar{\mathbf{M}}_{il} = \begin{bmatrix} \mathbf{Y}_{i1} + \mathbf{Y}_{i2} & -\bar{\mathbf{L}}_{il} \\ \mathbf{Y}_{i3} & \mathbf{0} \end{bmatrix}, \quad G_i = \begin{bmatrix} \mathbf{0} & \mathbf{0} \\ \star & \xi_i \lambda \mathbf{I} \end{bmatrix}$$

$$\bar{\mathbf{D}}_l = \begin{bmatrix} \mathbf{0} & \mathbf{F}_l \end{bmatrix}, \quad \bar{\mathbf{N}} = \begin{bmatrix} \mathbf{0} & \mathbf{I} \end{bmatrix}, \quad \lambda = \epsilon_1^2 + \epsilon_2^2 + \epsilon_3^2 + \epsilon_4^2 + \epsilon_5^2 + \epsilon_6^2.$$

Moreover, the observer gains are determined by $\mathbf{L}_{il} = \mathbf{Y}_{i1}^{-1}\bar{\mathbf{L}}_{il}, l \in \Gamma(i), i \in \ell$.

Proof. The proof of this theorem follows a similar line as Theorem 6.2 and thus is omitted here for simplification. $\qquad\square$

Consequently, the following scheme is employed for the realization of a robust FD system for discrete-time nonlinear processes (6.68):

- Run the residual generator (6.35)

- Set

$$\bar{J}_{\text{th}} = \|\mathbf{u}_\tau\|_2^2 + \delta_{\mathbf{d}}^2 + \sup_{\mathbf{x}(0), \hat{\mathbf{x}}(0)} \bar{\mathbf{x}}^T(0)\bar{\mathcal{P}}_{i_o}\bar{\mathbf{x}}(0) \tag{6.81}$$

- Set the detection logic

$$\begin{cases} \bar{J} = \|\mathbf{r}_\tau\|_2^2 > \bar{J}_{\text{th}} \implies \text{faulty} \\ \bar{J} = \|\mathbf{r}_\tau\|_2^2 \leq \bar{J}_{\text{th}} \implies \text{fault-free.} \end{cases} \tag{6.82}$$

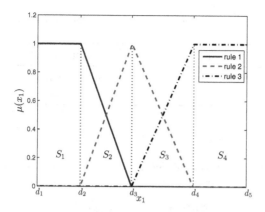

Figure 6.1: Membership functions for fuzzy modelling

6.4 A Numerical Example

Consider the following discrete-time T-S fuzzy dynamic system with three rules:

Plant rule $\Re^l$: *IF* $x_1(k)$ *is* N_1^l, *THEN*

$$\begin{cases} \mathbf{x}(k+1) = \mathbf{A}_l x(k) + \Delta_{\mathbf{A}}(\mathbf{x}(k), u(k))\mathbf{x}(k) + \mathbf{B}_l u(k) \\ \qquad + \Delta_{\mathbf{B}}(\mathbf{x}(k), u(k))u(k) \\ \mathbf{y}(k) = \mathbf{C}_l \mathbf{x}(k), \quad l \in \{1, 2, 3\} \end{cases}$$

where

$$\mathbf{A}_1 = \begin{bmatrix} 0.402 & -0.10 \\ 0.10 & 0.501 \end{bmatrix}, \ \mathbf{A}_2 = \begin{bmatrix} 0.882 & -0.018 \\ 0.018 & 0.54 \end{bmatrix}$$

$$\mathbf{A}_3 = \begin{bmatrix} 0.585 & -0.054 \\ 0.054 & 0.855 \end{bmatrix}, \mathbf{B}_1 = \begin{bmatrix} 0 \\ 0.1 \end{bmatrix}, \ \mathbf{B}_2 = \begin{bmatrix} 0 \\ 0.25 \end{bmatrix}$$

$$\mathbf{B}_3 = \begin{bmatrix} 0 \\ 0.2 \end{bmatrix}, \ \mathbf{C}_1 = \mathbf{C}_2 = \mathbf{C}_3 = \begin{bmatrix} 1 & 0 \end{bmatrix}.$$

It is assumed that the model uncertainties are bounded by $\|\Delta_{\mathbf{A}}(\mathbf{x}, u)\| \le 0.1, \|\Delta_{\mathbf{B}}(\mathbf{x}, u)\| \le 0.045$. The normalized membership functions for premise variable $x_1(k)$ are given in Fig. 6.1. Then, according to the

Table 6.1: Comparison of FD performance for different approaches

Approach	Performance
Common approach	$\alpha^2 = 0.473$
Piecewise-fuzzy approach	$\alpha^2 = 0.223$

partition method proposed in Section 6.1.2, the premise-variable space can be divided into four regions (see Fig. 6.1), which are given by

$$S_1 := \{x_1 | d_1 < x_1 \leq d_2\}$$
$$S_2 := \{x_1 | d_2 < x_1 \leq d_3\}$$
$$S_3 := \{x_1 | d_3 < x_1 \leq d_4\}$$
$$S_4 := \{x_1 | d_4 < x_1 \leq d_5\}$$

with $d_1 = -\infty$, $d_2 = -10$, $d_3 = 0$, $d_4 = 15$, $d_5 = \infty$.

The major objective of this study is to show the effectiveness of the $\mathcal{L}_2$ observer-based FD system proposed in Section 6.1.2. The advantage of the piecewise-fuzzy Lypunov function-based approach over the common Lyapunov function-based approach is demonstrated first. For ease of presentation, residual signals of each region are weighted constantly first, that is, we first choose $w_i \equiv w = \alpha^{-1}, i = 1, 2, 3, 4$. As a result, the comparison between the performance of these two approaches is shown in Table 6.1. It becomes evident that by adopting piecewise-fuzzy Lyapunov function based approach, the most possible increment in fault detectability would be more than 50 %.

Then, a weighted piecewise-fuzzy observer-based residual generator of the form (6.35) is developed based on the obtained results in Theorem 6.2 first. Since $\mathcal{L}_2$ norm is used for the evaluation of residual signal, the weighting factors are computed by adopting Algorithm 3, which results in

$$\omega_1 = 2.626, \ \omega_2 = 2.137, \ \omega_3 = 2.119, \ \omega_4 = 2.209. \tag{6.83}$$

It turns out that

$$\omega_3 = \min\{\omega_i, \ i = 1, \cdots, 4\} = \alpha^{-1}.$$

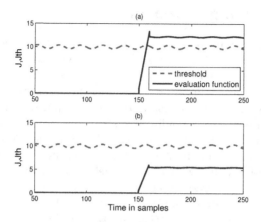

Figure 6.2: (a) Fault detection performance of FD system with weighting factors in (6.83); (b) fault detection performance of FD system with constant weighting factor $\omega = 2.119$

Thus, the FD system (6.35), (6.66) and (6.67) promises a higher fault detectability than the original FD system with a constant weighting factor. The most possible increment in fault detectability would be more than 35 %.

Next, the application of the above algorithm in improving fault detection performance is demonstrated. To this end, the input is chosen as $u(k) = 1 + 0.03\sin(3k)$ and an additive fault $f(k) = 0.72, k \geq 150$ is simulated in the measurement model. By setting the evaluation window as $\tau = 10$ samples, the comparison result of the fault detection performance of FD system with weighting factors (6.83) and FD system with a constant weighting factor $\omega = 2.119$ is shown in Fig 6.2. It can be evidently seen that the FD scheme with weighting factors (6.83) is more sensitive to the fault than the traditional constantly weighted FD system. Thus, it can be concluded that the fault detectability is significantly improved by weighting the residual signal of each region by means of different weighting factors.

6.5 Concluding Remarks

In this chapter, the integrated $\mathcal{L}_2$ observer-based FD systems, composed by an observer-based residual generator, an evaluation function and a decision logic with an embedded threshold, have been studied for a general type of discrete-time nonlinear processes. To gain a deeper insight into this FD framework, the generalization of the weak output re-constructability condition to discrete-time nonlinear systems has been introduced first, which has been proved to be the existence condition for an $\mathcal{L}_2$ observer-based FD system. Then, an integrated design scheme has been proposed with the aid of T-S fuzzy dynamic modelling technique. Specifically, a weighted piecewise-fuzzy observer-based residual generator has been proposed aiming at improving the fault detectability of conventional fuzzy observer-based FD approaches. All the solutions are derived based on piecewise-fuzzy Lyapunov functions in terms of LMIs. The basic idea behind the proposed scheme is to weight the residual signals of each local region individually, instead of constantly, by means of the weighting factors. This is strongly motivated by the fact that the local behavior of nonlinear systems can be considerably different.

7 FTC Configurations for Nonlinear Systems

Youla parametrization plays an essential role in analyzing and optimizing control system performance for LTI systems [139, 170]. Numerous controller architectures, developed from Youla parametrization, have been proposed over the past decades [129, 171, 34]. In recent years, some researchers have made the effort to generalize Youla parametrization to nonlinear systems with the aid of nonlinear factorization techniques [130, 104, 136, 105, 24]. In particular, all the stabilizing plant-controller pairs for nonlinear systems have been studied in [105]. Due to the dependence on the initial conditions of the plant, its implementation is limited. To solve this problem, an alternative realization in terms of detectable SKR has been formulated in [46, 44]. Meanwhile, observer-based realizations of the controller parameterization have been studied in a series of works [56, 9, 91].

The main objective of this chapter lies in residual generator based FTC configurations. Proceeding from the results given in [46], the observer and residual generator based realization of controller parametrization for affine nonlinear systems is first proposed, which is composed by an observer-based nominal controller and a residual-driven dynamic compensator. It allows us to attain the controller modification without interrupting runtime operations by only adjusting the dynamic compensator. Regarding the high demands for system reliability, a fault diagnosis system is further integrated into the residual generator based controller to construct the FTC configuration. Furthermore, a new interpretation for fault-tolerant controller design is revealed with any stabilizing controller as the nominal controller. Another contribution of this chapter is one integrated design scheme for fault diagnosis system, nominal controller and residual-driven dynamic compensator for the FTC configuration. Moreover, the fault diagnosis system in terms of the parameterized residual generators is specifically discussed. It is noteworthy that the proposed scheme will

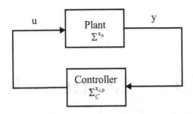

Figure 7.1: Standard feedback control loop

be suitable for the maintenance and life-circle management for affine nonlinear systems.

7.1 Preliminaries and Problem Formulation

Consider the standard feedback control loop sketched in Fig. 7.1. Suppose that $\Sigma^{\mathbf{x}_0}$ is an affine nonlinear system described by

$$\Sigma^{\mathbf{x}_0} : \begin{cases} \dot{\mathbf{x}} = \mathbf{a}(\mathbf{x}) + \mathbf{b}(\mathbf{x})\mathbf{u} \\ \mathbf{y} = \mathbf{c}(\mathbf{x}) \end{cases} \tag{7.1}$$

where $\mathbf{x}$, $\mathbf{u}$ and $\mathbf{y}$ denote the state, input and output vectors respectively. $\mathbf{x}_0 \in \mathcal{X}^0$ denotes the initial condition. $\mathbf{a}(\mathbf{x})$, $\mathbf{b}(\mathbf{x})$ and $\mathbf{c}(\mathbf{x})$ are smooth nonlinear functions with appropriate dimensions.

In this chapter, the FTC configurations for nonlinear systems (7.1) will be investigated. For this purpose, the observer-based controller parametrization forms for affine nonlinear systems are proposed first. Then the FTC configurations will be constructed by virtue of the proposed controller structures and an embedded fault diagnosis system. The further task is dedicated to propose a design scheme and apply it to a numerical example.

7.2 Residual Generator-based FTC Configurations

This section aims to investigate the EIMC structures for affine nonlinear systems (7.1), and further develop the FTC configurations.

7.2.1 Controller Parametrization

Before proceeding further, the existing results on controller parametrization for nonlinear systems by means of input-output approach (operators) are recalled first.

It is well-known that system (7.1) under consideration can be described by an operator

$$\Sigma^{\mathbf{x}_0} : \mathcal{U} \to \mathcal{Y}. \tag{7.2}$$

It follows from Definition 3.8 that an SKR for plant $\Sigma^{\mathbf{x}_0}$ with the same initial condition can be denoted by

$$K_\Sigma^{\mathbf{x}_0} : \mathcal{U} \times \mathcal{Y} \to \mathcal{Z}_\Sigma. \tag{7.3}$$

For our purpose, an SKR with different initial conditions for plant $\Sigma^{\mathbf{x}_0}$ is expressed as

$$K_{\bar{\Sigma}}^{\bar{\mathbf{x}}_0} : \mathcal{U} \times \mathcal{Y} \to \mathcal{Z}_{\bar{\Sigma}}. \tag{7.4}$$

Definition 7.1. *An SKR $K_{\bar{\Sigma}}^{\bar{\mathbf{x}}_0}$ of $\Sigma^{\mathbf{x}_0}$ is said to be detectable, if for $\forall \bar{\mathbf{x}}_0 \in \mathcal{X}^0$ and $\forall (\mathbf{u}, \mathbf{y}) \in \mathcal{U}^s \times \mathcal{Y}^s$*

$$K_{\bar{\Sigma}}^{\bar{\mathbf{x}}_0}(\mathbf{u}, \mathbf{y}) - K_\Sigma^{\mathbf{x}_0}(\mathbf{u}, \mathbf{y}) \in \mathcal{E}^s. \tag{7.5}$$

For simplicity, (7.3) and (7.4) are represented as $\mathbf{z}_\Sigma = K_\Sigma^{\mathbf{x}_0}(\mathbf{u}, \mathbf{y})$ and $\mathbf{z}_{\bar{\Sigma}} = K_{\bar{\Sigma}}^{\bar{\mathbf{x}}_0}(\mathbf{u}, \mathbf{y})$, respectively. By considering $\mathcal{Z}_\Sigma$ as the signal space of residual signal $\mathbf{z}_\Sigma$, the definition of a well-defined SKR can be summarized as follows.

Definition 7.2. *An SKR (7.3) of (7.2) is said to be well-defined, if there exists a pseudo-inverse system $K_\Sigma^\sharp : \mathcal{U} \times \mathcal{Z}_\Sigma \to \mathcal{Y}$ such that*

$$\mathbf{y} = K_\Sigma^\sharp(\mathbf{u}, \mathbf{z}_\Sigma) \iff K_\Sigma^{\mathbf{x}_0}(\mathbf{u}, \mathbf{y}) = \mathbf{z}_\Sigma \tag{7.6}$$

holds for $\forall \mathbf{x}_0 \in \hat{\mathcal{X}}^0$, $\forall \mathbf{u} \in \mathcal{U}$, $\forall \mathbf{y} \in \mathcal{Y}$ and $\mathbf{z}_\Sigma \in \mathcal{Z}_\Sigma$.

Consider the feedback control loop shown in Fig. 7.1, where $\Sigma^{\mathbf{x}_0} : \mathcal{U} \to \mathcal{Y}$ is the nonlinear plant and $\Sigma_C^{\mathbf{x}_{c,0}} : \mathcal{Y} \to \mathcal{U}$ is the feedback controller with the following SKRs

$$K_\Sigma^{\mathbf{x}_0} : \mathcal{U} \times \mathcal{Y} \to \mathcal{Z}_\Sigma \tag{7.7}$$

$$K_{\Sigma_C}^{\mathbf{x}_{c,0}} : \mathcal{Y} \times \mathcal{U} \to \mathcal{Z}_{\Sigma_C} \tag{7.8}$$

respectively. Then the SKR of the closed-loop system $\{\Sigma^{\mathbf{x}_0}, \Sigma_C^{\mathbf{x}_{c,0}}\}$ is defined as follows:

$$K_{\{\Sigma,\Sigma_C\}}^{\mathbf{x}_0,\mathbf{x}_{c,0}} : \mathcal{U} \times \mathcal{Y} \to \mathcal{Z}_\Sigma \times \mathcal{Z}_{\Sigma_C}$$

$$\begin{pmatrix} \mathbf{u} \\ \mathbf{y} \end{pmatrix} \mapsto \begin{pmatrix} \mathbf{z}_\Sigma \\ \mathbf{z}_{\Sigma_C} \end{pmatrix} = \begin{pmatrix} K_\Sigma^{\mathbf{x}_0}(\mathbf{u},\mathbf{y}) \\ K_{\Sigma_C}^{\mathbf{x}_{c,0}}(\mathbf{y},\mathbf{u}) \end{pmatrix}. \tag{7.9}$$

As a result, the stability of the closed-loop system can be defined in terms of SKR (7.9).

Definition 7.3. *[46] The closed-loop system $\{\Sigma^{\mathbf{x}_0}, \Sigma_C^{\mathbf{x}_{c,0}}\}$ is said to be internally stable, if for $\forall \mathbf{x}_0 \in \mathcal{X}^0$ and $\forall \mathbf{x}_{c,0} \in \mathcal{X}_C^0$, $K_{\{\Sigma,\Sigma_C\}}^{-1}$ is stable.*

The existing result on controller parametrization for nonlinear systems in terms of nonlinear operators is recalled first (see Fig. 7.2).

Lemma 7.1. *[46] Consider an internally stable system $\{\Sigma^{\mathbf{x}_0}, \Sigma_C^{\mathbf{x}_{c,0}}\}$ with an SKR (7.9). Suppose $\Sigma_Q^{\mathbf{x}_{q,0}}$ is any system with a well-defined SKR*

$$K_{\Sigma_Q}^{\mathbf{x}_{q,0}} : \mathcal{Z}_\Sigma \times \mathcal{Z}_{\Sigma_C} \to \mathcal{Z}_{\Sigma_Q}. \tag{7.10}$$

Suppose that $K_\Sigma^{\mathbf{x}_0}$ is detectable, if the controller $\Sigma_C^{\mathbf{x}_{c,0}}$ has the following well-defined SKR:

$$K_{\Sigma_{C_Q}}^{\bar{\mathbf{x}}_0,\mathbf{x}_{q,0}} = K_{\Sigma_Q}^{\mathbf{x}_{q,0}} \circ K_{\{\bar{\Sigma},\Sigma_C\}}^{\bar{\mathbf{x}}_0,\mathbf{x}_{c,0}} \tag{7.11}$$

then the closed-loop system $\{\Sigma^{\mathbf{x}_0}, \Sigma_{C_Q}^{\mathbf{x}_{c,0}}\}$ with SKR $K_{\{\Sigma,\Sigma_{C_Q}\}}^{\mathbf{x}_0,\mathbf{x}_{c,0},\mathbf{x}_{q,0}}$ is internally stable if and only if for $\forall \mathbf{x}_0 \in \mathcal{X}^0$, $K_{\{\Sigma,\Sigma_{C_Q}\}}^{\mathbf{x}_0,\mathbf{x}_{c,0},\mathbf{x}_{q,0}}$ is invertible and $K_{\Sigma_Q}^{\sharp}$ is stable.

Furthermore, given an internally stable system $\{\Sigma^{\mathbf{x}_0}, \Sigma_{C^*}^{\mathbf{x}_{c^*,0}}\}$ with an SKR $K_{\{\Sigma,\Sigma_{C^*}\}}^{\mathbf{x}_0,\mathbf{x}_{c^*,0}}$, there exists a well-defined kernel representation $K_{\Sigma_{Q^*}}^{\mathbf{x}_{q^*,0}}$ such that $K_{\Sigma_{C_{Q^*}}}^{\bar{\mathbf{x}}_0,\mathbf{x}_{q^*,0}} = K_{\Sigma_{C^*}}^{\mathbf{x}_{c^*,0}}$ holds and $K_{\Sigma_{Q^*}}^{\sharp}$ is stable.

Remark 7.1. *Note that if the initial condition of the plant is available, an alternative controller parametrization form has been proposed in [105]. However, the dependence on the initial condition limits its application in practice. To this end, the observer-based controller parametrization form (EIMC structure) will be investigated for affine nonlinear systems by means of detectable kernel representation proposed in Lemma 7.1 in the subsequent sections.*

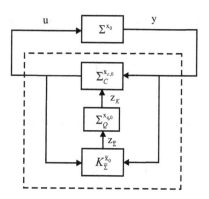

Figure 7.2: Controller parametrization via detectable kernel representations

7.2.2 Residual Generator-based Realizations of Controller Parameterization

This section presents the observer-based realizations of controller parameterization for nonlinear systems (7.1). It follows from [134] that one realization of SKR for $\Sigma^{\mathbf{x}_0}$ with the same initial condition is of the following form:

$$K_\Sigma^{\mathbf{x}_0} : \begin{cases} \dot{\mathbf{x}} = \mathbf{a}(\mathbf{x}) + \mathbf{b}(\mathbf{x})\mathbf{u} + \mathbf{l}(\mathbf{x})(\mathbf{y} - \mathbf{c}(\mathbf{x})) \\ \mathbf{z}_\Sigma = \mathbf{y} - \mathbf{c}(\mathbf{x}) \end{cases} \tag{7.12}$$

where $\mathbf{l}(\mathbf{x})$ denotes the gain matrix which will be determined later. To avoid the dependence on the initial conditions, as mentioned above, $K_{\bar{\Sigma}}^{\bar{\mathbf{x}}_0}$, as a copy of $K_\Sigma^{\mathbf{x}_0}$ with different initial condition, can be constructed as follows:

$$K_{\bar{\Sigma}}^{\bar{\mathbf{x}}_0} : \begin{cases} \dot{\bar{\mathbf{x}}} = \mathbf{a}(\bar{\mathbf{x}}) + \mathbf{b}(\bar{\mathbf{x}})\mathbf{u} + \mathbf{l}(\bar{\mathbf{x}})(\mathbf{y} - \mathbf{c}(\bar{\mathbf{x}})) \\ \mathbf{z}_{\bar{\Sigma}} = \mathbf{y} - \mathbf{c}(\bar{\mathbf{x}}). \end{cases} \tag{7.13}$$

Thus, $\mathbf{z}_{\bar{\Sigma}}$ provides an estimate of $\mathbf{z}_\Sigma$, which contains the most important information needed for fault detection. Therefore, the SKR $K_{\bar{\Sigma}}^{\bar{\mathbf{x}}_0}$ can be considered as a residual generator, with $\mathbf{z}_{\bar{\Sigma}}$ as the residual signal and $\bar{\mathbf{x}}$ as the state estimation. The residual generator considered here is a key component as it continuously produces state estimates and residual signals.

Without loss of generality, the state-space representation of controller $\Sigma_C^{\mathbf{x}_{c,0}}$ can be given by

$$\Sigma_C^{\mathbf{x}_{c,0}} : \begin{cases} \dot{\mathbf{x}}_c = \mathbf{f}_c(\mathbf{x}_c, \mathbf{y}) \\ \mathbf{u} = \mathbf{h}_c(\mathbf{x}_c, \mathbf{y}). \end{cases} \tag{7.14}$$

The SKR can be constructed of the following form:

$$K_{\Sigma_C}^{\hat{\mathbf{x}}_{c,0}} : \begin{cases} \dot{\hat{\mathbf{x}}}_c = \boldsymbol{\phi}_c(\hat{\mathbf{x}}_c, \mathbf{u}, \mathbf{y}) \\ \mathbf{z}_{\Sigma_C} = \boldsymbol{\varphi}_c(\hat{\mathbf{x}}_c, \mathbf{u}, \mathbf{y}). \end{cases} \tag{7.15}$$

In the sequel, the observer-based realization of the controller parametrization form (7.11) is presented, which plays an essential role in formulating the FTC configurations.

Theorem 7.1. *[84] Given the feedback control system shown in Fig. 7.1, where $\Sigma^{\mathbf{x}_0}$ is affine nonlinear system (7.1). Suppose $\Sigma^{\mathbf{x}_0}$ has a detectable SKR (7.13) and the following nominal controller*

$$\Sigma_{C_n}^{\bar{\mathbf{x}}_0} : \begin{cases} \dot{\bar{\mathbf{x}}} = \mathbf{a}(\bar{\mathbf{x}}) + \mathbf{b}(\bar{\mathbf{x}})\mathbf{k}(\bar{\mathbf{x}}) + \mathbf{l}(\bar{\mathbf{x}})(\mathbf{y} - \mathbf{c}(\bar{\mathbf{x}})) \\ \mathbf{u} = \mathbf{k}(\bar{\mathbf{x}}) \end{cases} \tag{7.16}$$

internally stabilize the plant, then all the controllers that internally stabilize the plant can be parameterized by

$$\Sigma_{C_Q}^{\bar{\mathbf{x}}_0, \mathbf{x}_{q,0}} : \begin{cases} \dot{\bar{\mathbf{x}}} = \mathbf{a}(\bar{\mathbf{x}}) + \mathbf{b}(\bar{\mathbf{x}})\mathbf{u} + \mathbf{l}(\bar{\mathbf{x}})(\mathbf{y} - \mathbf{c}(\bar{\mathbf{x}})) \\ \mathbf{z}_{\bar{\Sigma}} = \mathbf{y} - \mathbf{c}(\bar{\mathbf{x}}) \\ \dot{\mathbf{x}}_q = \mathbf{f}_q(\mathbf{x}_q, \mathbf{z}_{\bar{\Sigma}}) \\ \mathbf{u} = \mathbf{k}(\bar{\mathbf{x}}) + \mathbf{h}_q(\mathbf{x}_q, \mathbf{z}_{\bar{\Sigma}}) \end{cases} \tag{7.17}$$

where $\Sigma_Q^{\mathbf{x}_{q,0}}$ is any system

$$\Sigma_Q^{\mathbf{x}_{q,0}} : \begin{cases} \dot{\mathbf{x}}_q = \mathbf{f}_q(\mathbf{x}_q, \mathbf{z}_{\bar{\Sigma}}) \\ \mathbf{z}_K = \mathbf{h}_q(\mathbf{x}_q, \mathbf{z}_{\bar{\Sigma}}) \end{cases} \tag{7.18}$$

with a well-defined SKR and stable $K_{\Sigma_Q}^{\sharp}$.

Proof. According to Fig. 7.2, $K_{\bar{\Sigma}}^{\bar{\mathbf{x}}_0}$ is a detectable SKR for $\Sigma^{\mathbf{x}_0}$ with the state-space representation

$$K_{\bar{\Sigma}}^{\bar{\mathbf{x}}_0} : \begin{cases} \dot{\bar{\mathbf{x}}} = \mathbf{a}(\bar{\mathbf{x}}) + \mathbf{b}(\bar{\mathbf{x}})\mathbf{u} + \mathbf{l}(\bar{\mathbf{x}})(\mathbf{y} - \mathbf{c}(\bar{\mathbf{x}})) \\ \mathbf{z}_{\bar{\Sigma}} = \mathbf{y} - \mathbf{c}(\bar{\mathbf{x}}). \end{cases} \tag{7.19}$$

$\Sigma_Q^{\mathbf{x}_{q,0}}$ is an operator with $\mathbf{z}_{\bar{\Sigma}}$ as input and $\mathbf{z}_K$ as output, which can be represented as

$$\Sigma_Q^{\mathbf{x}_{q,0}} : \begin{cases} \dot{\mathbf{x}}_q = \mathbf{f}_q(\mathbf{x}_q, \mathbf{z}_{\bar{\Sigma}}) \\ \mathbf{z}_K = \mathbf{h}_q(\mathbf{x}_q, \mathbf{z}_{\bar{\Sigma}}). \end{cases} \tag{7.20}$$

Suppose $\Sigma_Q^{\mathbf{x}_{q,0}}$ has a well-defined SKR

$$K_{\Sigma_Q}^{\mathbf{x}_{q,0}} : \begin{cases} \dot{\mathbf{x}}_q = \mathbf{f}_q(\mathbf{x}_q, \mathbf{z}_{\bar{\Sigma}}) + \mathbf{l}_q(\mathbf{x}_q)(\mathbf{z}_K - \mathbf{h}_q(\mathbf{x}_q, \mathbf{z}_{\bar{\Sigma}})) \\ \mathbf{z}_Q = \mathbf{z}_K - \mathbf{h}_q(\mathbf{x}_q, \mathbf{z}_{\bar{\Sigma}}) \end{cases} \tag{7.21}$$

where $\mathbf{l}_q(\cdot)$ is designed to guarantee the stability of $K_{\Sigma_Q}^{\mathbf{x}_{q,0}}$.

It is straightforward to obtain the following SKR for $\Sigma_{C_n}^{\bar{\mathbf{x}}_0}$

$$K_{\Sigma_{C_n}}^{\bar{\mathbf{x}}_0} : \begin{cases} \dot{\bar{\mathbf{x}}} = \mathbf{a}(\bar{\mathbf{x}}) + \mathbf{b}(\bar{\mathbf{x}})\mathbf{k}(\bar{\mathbf{x}}) + \mathbf{l}(\bar{\mathbf{x}})(\mathbf{y} - \mathbf{c}(\bar{\mathbf{x}})) + \mathbf{b}(\bar{\mathbf{x}})(\mathbf{u} - \mathbf{k}(\bar{\mathbf{x}})) \\ \mathbf{z}_K = \mathbf{u} - \mathbf{k}(\bar{\mathbf{x}}). \end{cases} \tag{7.22}$$

Moreover, it follows from Lemma 7.1 that

$$K_{\Sigma_{C_Q}}^{\bar{\mathbf{x}}_0, \mathbf{x}_{q,0}} : \begin{cases} \dot{\bar{\mathbf{x}}} = \mathbf{a}(\bar{\mathbf{x}}) - \mathbf{l}(\bar{\mathbf{x}})\mathbf{c}(\bar{\mathbf{x}}) + \mathbf{b}(\bar{\mathbf{x}})\mathbf{u} + \mathbf{l}(\bar{\mathbf{x}})\mathbf{y} \\ \mathbf{z}_{\bar{\Sigma}} = \mathbf{y} - \mathbf{c}(\bar{\mathbf{x}}) \\ \dot{\mathbf{x}}_q = \mathbf{f}_q(\mathbf{x}_q, \mathbf{z}_{\bar{\Sigma}}) + \mathbf{l}_q(\mathbf{x}_q)(\mathbf{z}_K - \mathbf{h}_q(\mathbf{x}_q, \mathbf{z}_{\bar{\Sigma}})) \\ \mathbf{z}_Q = \mathbf{u} - \mathbf{k}(\bar{\mathbf{x}}) - \mathbf{h}_q(\mathbf{x}_q, \mathbf{z}_{\bar{\Sigma}}). \end{cases} \tag{7.23}$$

For the sake of brevity, we set $\mathbf{z}_Q = \mathbf{0}$. It turns out that

$$K_{\Sigma_{C_Q}}^{\bar{\mathbf{x}}_0, \mathbf{x}_{q,0}} : \begin{cases} \dot{\bar{\mathbf{x}}} = \mathbf{a}(\bar{\mathbf{x}}) - \mathbf{l}(\bar{\mathbf{x}})\mathbf{c}(\bar{\mathbf{x}}) + \mathbf{b}(\bar{\mathbf{x}})\mathbf{u} + \mathbf{l}(\bar{\mathbf{x}})\mathbf{y} \\ \mathbf{z}_{\bar{\Sigma}} = \mathbf{y} - \mathbf{c}(\bar{\mathbf{x}}) \\ \dot{\mathbf{x}}_q = \mathbf{f}_q(\mathbf{x}_q, \mathbf{z}_{\bar{\Sigma}}) \\ \mathbf{0} = \mathbf{u} - \mathbf{k}(\bar{\mathbf{x}}) - \mathbf{h}_q(\mathbf{x}_q, \mathbf{z}_{\bar{\Sigma}}). \end{cases} \tag{7.24}$$

Then it is easy to verify that (7.24) is an SKR of the following system:

$$\Sigma_{C_Q}^{\bar{\mathbf{x}}_0, \mathbf{x}_{q,0}} : \begin{cases} \dot{\bar{\mathbf{x}}} = \mathbf{a}(\bar{\mathbf{x}}) - \mathbf{l}(\bar{\mathbf{x}})\mathbf{c}(\bar{\mathbf{x}}) + \mathbf{b}(\bar{\mathbf{x}})\mathbf{k}(\bar{\mathbf{x}}) + \mathbf{l}(\bar{\mathbf{x}})\mathbf{y} + \mathbf{b}(\bar{\mathbf{x}})\mathbf{h}_q(\mathbf{x}_q, \mathbf{z}_{\bar{\Sigma}}) \\ \mathbf{z}_{\bar{\Sigma}} = \mathbf{y} - \mathbf{c}(\bar{\mathbf{x}}) \\ \dot{\mathbf{x}}_q = \mathbf{f}_q(\mathbf{x}_q, \mathbf{z}_{\bar{\Sigma}}) \\ \mathbf{u} = \mathbf{k}(\bar{\mathbf{x}}) + \mathbf{h}_q(\mathbf{x}_q, \mathbf{z}_{\bar{\Sigma}}). \end{cases}$$

$$(7.25)$$

Thus, the controller scheme (7.17) is obtained.

Recall that $\Sigma_C^{\bar{\mathbf{x}}_0}$ can internally stabilize the plant, then (7.17) can internally stabilize the closed loop if and only if $K_{\{\Sigma, \Sigma_{C_Q}\}}^{\mathbf{x}_0, \mathbf{x}_{q,0}}$ is invertible and $K_{\Sigma_Q}^{\sharp}$ is stable. It can be easily verified that $K_{\{\Sigma, \Sigma_{C_Q}\}}^{\mathbf{x}_0, \mathbf{x}_{q,0}}$ is always invertible in this case. Therefore, (7.17) can internally stabilize the closed loop as long as $K_{\Sigma_Q}^{\sharp}$ is stable.

According to Lemma 7.1, for any controller $\Sigma_{C^*}^{\mathbf{x}_0^*}$ that can internally stabilize the feedback system in Fig. 7.1, there exists a well-defined SKR $K_{\Sigma_{Q^*}}^{\mathbf{x}_{q^*,0}}$, such that $K_{\Sigma_{C_{Q^*}}}^{\bar{\mathbf{x}}_0, \mathbf{x}_{q^*,0}} = K_{\Sigma_{C^*}}^{\mathbf{x}_0^*}$ holds. As all the controller $\Sigma_{C_{Q^*}}^{\bar{\mathbf{x}}_0, \mathbf{x}_{q^*,0}}$ can be realized in terms of (7.17), it is seen that (7.17) can represent all the stabilizing controllers for plant $\Sigma^{\mathbf{x}_0}$. The proof is thus completed. $\square$

Remark 7.2. *It is worth mentioning that the nominal controller (7.16) can be any observer-based controller that internally stabilize the plant (7.1).*

Remark 7.3. *It is noted that the controller (7.17) can be considered as the nonlinear extension of the EIMC structure proposed by [34]. The controller parametrization enables the stabilization of the plant and the optimization of the performance criteria within the class. It is composed of two separate modules: an observer-based nominal controller $\Sigma_{C_n}^{\bar{\mathbf{x}}_0}$ and a residual-driven compensator $\Sigma_Q^{\mathbf{x}_{q,0}}$. $\Sigma_{C_n}^{\bar{\mathbf{x}}_0}$ is designed to deliver the system performance. $\Sigma_Q^{\mathbf{x}_{q,0}}$, as an additional parameter, provides the potential to enhance the reliability of the system, which will be discussed later.*

Furthermore, the following theorem is put forward to reveal a new representation of the controller configuration (7.17) and the distinguished role of the residual signal.

Theorem 7.2. *Given the feedback control loop (in Fig. 7.1) with an internally stabilizing controller $\Sigma_{C_0}^{\mathbf{x}_{c_0,0}} : \mathcal{Y} \to \mathcal{U}$, then all the controllers* *that internally stabilize the control loop can be parameterized by*

$$\mathbf{u} = \Sigma_{C_0}^{\mathbf{x}_{c_0},0}(\mathbf{u},\mathbf{y}) + \Sigma_{Q_c}^{\mathbf{x}_{q_c},0} \circ K_{\bar{\Sigma}}^{\bar{\mathbf{x}}_0}(\mathbf{u},\mathbf{y}) \tag{7.26}$$

where $\Sigma_{Q_c}^{\mathbf{x}_{q_c},0}$ has a well-defined SKR such that $K_{\bar{\Sigma}_{Q_c}}^{\sharp}$ is stable.

Proof. Recall that (7.17) can span the space of stabilizing controllers for plant $\Sigma^{\mathbf{x}_0}$ by varying $\Sigma_Q^{\mathbf{x}_q,0}$. Note that (7.20) can be denoted by the cascade of two operators $\Sigma_Q^{\mathbf{x}_q,0}$ and $K_{\bar{\Sigma}}^{\bar{\mathbf{x}}_0}$ as $\mathbf{z}_K = \Sigma_Q^{\mathbf{x}_q,0} \circ K_{\bar{\Sigma}}^{\bar{\mathbf{x}}_0}(\mathbf{u},\mathbf{y})$ for simplicity. Thus for any internally stabilizing controller $\Sigma_{C_0}^{\mathbf{x}_{c_0},0}$, there exists a system $\Sigma_{Q_0}^{\mathbf{x}_{q_0},0} : \mathcal{Z}_{\bar{\Sigma}} \to \mathcal{U}$ such that

$$\Sigma_{C_0}^{\mathbf{x}_{c_0},0}(\mathbf{u},\mathbf{y}) = \Sigma_C^{\bar{\mathbf{x}}_0}(\mathbf{u},\mathbf{y}) + \Sigma_{Q_0}^{\mathbf{x}_{q_0},0} \circ K_{\bar{\Sigma}}^{\bar{\mathbf{x}}_0}(\mathbf{u},\mathbf{y}) \tag{7.27}$$

Furthermore, it is obvious that

$$\mathbf{u} = \Sigma_{C_0}^{\mathbf{x}_{c_0},0}(\mathbf{u},\mathbf{y}) + \Sigma_Q^{\mathbf{x}_q,0} \circ K_{\bar{\Sigma}}^{\bar{\mathbf{x}}_0}(\mathbf{u},\mathbf{y}) - \Sigma_{Q_0}^{\mathbf{x}_{q_0},0} \circ K_{\bar{\Sigma}}^{\bar{\mathbf{x}}_0}(\mathbf{u},\mathbf{y}) \tag{7.28}$$

By denoting

$$\Sigma_{Q_d}^{\mathbf{x}_{q_d},0} \circ K_{\bar{\Sigma}}^{\bar{\mathbf{x}}_0}(\mathbf{u},\mathbf{y}) = \Sigma_Q^{\mathbf{x}_q,0} \circ K_{\bar{\Sigma}}^{\bar{\mathbf{x}}_0}(\mathbf{u},\mathbf{y}) - \Sigma_{Q_0}^{\mathbf{x}_{q_0},0} \circ K_{\bar{\Sigma}}^{\bar{\mathbf{x}}_0}(\mathbf{u},\mathbf{y}) \tag{7.29}$$

one has that

$$\mathbf{u} = \Sigma_{C_0}^{\mathbf{x}_{c_0},0}(\mathbf{u},\mathbf{y}) + \Sigma_{Q_d}^{\mathbf{x}_{q_d},0} \circ K_{\bar{\Sigma}}^{\bar{\mathbf{x}}_0}(\mathbf{u},\mathbf{y}) \tag{7.30}$$

Therefore, for any internally stabilizing controller $\Sigma_{C_0}^{\mathbf{x}_{c_0},0} : \mathcal{Y} \to \mathcal{U}$, the set of all stabilizing controllers for the affine plant can be parameterized by (7.26). The proof is thus completed. $\qquad\square$

Remark 7.4. *Here, the most remarkable difference from controller structure (7.17) is any internally stabilizing controller $\Sigma_{C_0}^{\mathbf{x}_{c_0},0}$ can be chosen as the nominal controller, instead of the observer-based controller $\Sigma_{C_n}^{\bar{\mathbf{x}}_0}$. For instance, the widely applied state feedback controller, proportional-integral (PI) controller, proportional-integral-derivative (PID) controller and fuzzy output feedback controller [113] can also serve as the nominal controller. This is motivated from the viewpoint of practical applications.*

Similar to the controller parametrization form (7.17), the state-space representation of (7.26) can be written as follows:

$$\Sigma_{C_{Q_d}}^{\mathbf{x}_0,\mathbf{x}_{q_c},0} : \begin{cases} \dot{\bar{\mathbf{x}}} = \mathbf{a}(\bar{\mathbf{x}}) + \mathbf{b}(\bar{\mathbf{x}})\mathbf{u} + \mathbf{l}(\bar{\mathbf{x}})(\mathbf{y} - \mathbf{c}(\bar{\mathbf{x}})) \\ \dot{\mathbf{x}}_0 = \mathbf{f}_0(\mathbf{x}_0,\mathbf{y}) \\ \dot{\mathbf{x}}_{q_c} = \mathbf{f}_{q_c}(\mathbf{x}_{q_c},\mathbf{y} - \mathbf{c}(\bar{\mathbf{x}})) \\ \mathbf{u} = \mathbf{h}_0(\mathbf{x}_0,\mathbf{y}) + \mathbf{h}_{q_c}(\mathbf{x}_{q_c},\mathbf{y} - \mathbf{c}(\bar{\mathbf{x}})) \end{cases} \tag{7.31}$$

where $\Sigma_{C_0}^{\mathbf{x}_{c0},0}$ is any controller

$$\Sigma_{C_0}^{\mathbf{x}_{c0},0} : \begin{cases} \dot{\mathbf{x}}_0 = \mathbf{f}_0(\mathbf{x}_0, \mathbf{y}) \\ \mathbf{u}_0 = \mathbf{h}_0(\mathbf{x}_0, \mathbf{y}) \end{cases} \tag{7.32}$$

that can internally stabilize the plant (7.1) and $\Sigma_{Q_d}^{\mathbf{x}_{q_d},0}$ is any system

$$\Sigma_{Q_d}^{\mathbf{x}_{q_d},0} : \begin{cases} \dot{\mathbf{x}}_{q_d} = \mathbf{f}_{q_d}(\mathbf{x}_{q_d}, \mathbf{z}_{\bar{\Sigma}}) \\ \mathbf{z}_K = \mathbf{h}_{q_d}(\mathbf{x}_{q_d}, \mathbf{z}_{\bar{\Sigma}}) \end{cases} \tag{7.33}$$

with a well-defined SKR such that $K_{\Sigma_{Q_d}}^{\sharp}$ is stable.

7.2.3 FTC Configurations

In the previous section, all stabilizing controllers for affine nonlinear plant $\Sigma^{\mathbf{x}_0}$ are characterized in a residual generator based form. This section is devoted to develop two FTC configurations for affine nonlinear systems.

It has been observed that residual signal $\mathbf{z}_{\bar{\Sigma}}$ delivers the information of the fault signals. This fact establishes the relationship between control and diagnostics, and motivates us to integrate the design of feedback controllers with the setup of the residual signals. From this perspective, the free parameter $\Sigma_Q^{\mathbf{x}_q,0}$ in (7.17) can be utilized to compensate the fault-induced changes. Inspired by the above point, the FTC configuration is proposed based on (7.17), with the integration of a fault diagnosis system. Here, the fault diagnosis system is integrated to identify the fault-induced changes such that the fault-tolerant system can online accommodate the control law in response to the FD decisions.

On the other hand, it is worth mentioning that Theorem 3.4 provides us with the additional design freedom to obtain a better fault diagnosis performance in terms of nonlinear operators. By virtue of Theorem 7.1 and Theorem 3.4 , the following FTC proposition is formulated.

Proposition 4.1 *The FTC configuration depicted in Fig. 7.3 can realize an FTC for affine nonlinear systems by adopting the following rules:*

- *The SKR $K_{\bar{\Sigma}}^{\bar{\mathbf{x}}_0}$ is determined to provide the state estimation and residual signal.*

- *The fault diagnosis system is constructed by*

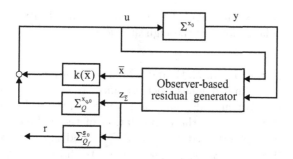

Figure 7.3: The FTC configuration

- *optimizing the fault diagnosis filter $\Sigma_{Q_f}^{s_0}$ to get the residual evaluation function, such as $J(\mathbf{r}) = \|\mathbf{r}\|_2^2$.*
- *determining the threshold J_{th}.*
- *formulating the decision logic:*

$$\begin{cases} J(\mathbf{r}) > J_{\text{th}} \Longrightarrow \text{Faulty} \\ J(\mathbf{r}) \leq J_{\text{th}} \Longrightarrow \text{Fault} - \text{free}. \end{cases}$$

- *The nominal controller $\Sigma_C^{\mathbf{x}_c,0}$ is designed to stabilize the plant as a priori.*

- *$\Sigma_Q^{\mathbf{x}_q,0}$ is triggered by the fault diagnosis system. If a fault alarm is released, $\Sigma_Q^{\mathbf{x}_q,0}$ will be activated to compensate the fault signal instead of reconfiguring the nominal controller.*

The above FTC configuration can be also considered as a lifecycle management scheme. The critical factor is the ability to respond to changing conditions efficiently, flexibly, and above all, with foresight. Along with the line of our previous work in [153], the system parameters are classified in the high and low priorities as follows:

- Recall that system stability is the minimum requirement towards industrial process control systems. This inspires us a real-time adaptation of $\mathbf{l}(\bar{\mathbf{x}})$ and $\mathbf{k}(\bar{\mathbf{x}})$ to potential changes in the system dynamics, and guarantee the overall system stability. For this reason,

$l(\bar{x})$ and $k(\bar{x})$ are labeled as H-PRIO (high-priority) parameters whose adaptation has the highest priority.

- On the other hand, $\Sigma_Q^{\mathbf{x}_q,0}$ and $\Sigma_{Q_f}^{\mathbf{s}_0}$ are employed to optimize the control and fault diagnosis performance. In case that a temporary system performance degradation is tolerable, the real-time demand and the priority for the optimization of these three parameters are relatively low. At this rate, $\Sigma_Q^{\mathbf{x}_q,0}$ and $\Sigma_{Q_f}^{\mathbf{s}_0}$ are considered as L-PRIO (low-priority) parameters. It allows us to develop an effective management scheme to deal with operational parameter or runtime configuration changes. The H-PRIO parameters are closely interrelated with the stability of the closed-loop system, while the L-PRIO parameters have no influence on system stability, but are essential for the optimization of system robustness, fault detectability and reference tracking performances.

Now, an alternative fault-tolerant configuration is addressed. The core of this structure is, in fact, the new interpretation for the residual generation based controller parametrization proposed in Theorem 7.2.

Proposition 4.2 *The FTC configuration in Fig. 7.4 can promise the same FTC ability for affine nonlinear systems as the one in Fig. 7.3.*

The free parameter $\Sigma_{Q_d}^{\mathbf{x}_{q_d},0}$ here provides us the potential to perform better than the conventional fault-tolerant techniques, for $\Sigma_{C_0}^{\mathbf{x}_{c_0},0}$ can be any pre-designed stabilizing controller. The activation of the fault tolerant compensator $\Sigma_{Q_d}^{\mathbf{x}_{q_d},0}$ relies on the decisions given by the fault diagnosis systems.

7.3 A Design Scheme of FTC Configuration

In this section, the design and implementation of the FTC configuration proposed in Proposition 4.1 is addressed. The design of observer-based FD system, the nominal controller and the residual-driven compensator will be discussed, respectively.

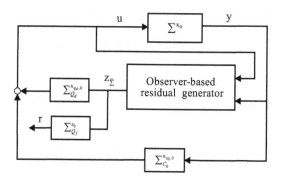

Figure 7.4: An alternative FTC configuration

7.3.1 Observer-based Fault Diagnosis System Design

At first, one design scheme for the $\mathcal{L}_2$ observer-based FD system is addressed. Denote

$$\begin{bmatrix} \dot{\mathbf{x}} \\ \dot{\bar{\mathbf{x}}} \end{bmatrix} = \bar{\mathbf{a}}(\mathbf{x}, \bar{\mathbf{x}}) + \bar{\mathbf{b}}(\mathbf{x}, \bar{\mathbf{x}})\mathbf{u} + \begin{bmatrix} 0 \\ \mathbf{l}(\bar{\mathbf{x}})\mathbf{z}_{\bar{\Sigma}} \end{bmatrix}$$

$$\bar{\mathbf{y}} = \mathbf{c}(\bar{\mathbf{x}}), \quad \mathbf{z}_{\bar{\Sigma}} = \mathbf{y} - \bar{\mathbf{y}}$$

$$\bar{\mathbf{a}}(\mathbf{x}, \bar{\mathbf{x}}) = \begin{bmatrix} \mathbf{a}(\mathbf{x}) \\ \mathbf{a}(\bar{\mathbf{x}}) \end{bmatrix}, \bar{\mathbf{b}}(\mathbf{x}, \bar{\mathbf{x}}) = \begin{bmatrix} \mathbf{b}(\mathbf{x}) \\ \mathbf{b}(\bar{\mathbf{x}}) \end{bmatrix}.$$

The determination of the gain matrix $\mathbf{l}(\bar{\mathbf{x}})$ of the SKR (7.13) will be addressed in the following theorem.

Theorem 7.3. *Given the system (7.1) and SKR (7.13). Suppose that there exists a constant $\gamma > 0$ such that there is a solution $V(\mathbf{x}, \bar{\mathbf{x}}) \geq 0$ to the following HJI*

$$V_{\mathbf{x},\bar{\mathbf{x}}}(\mathbf{x}, \bar{\mathbf{x}})\bar{\mathbf{a}}(\mathbf{x}, \bar{\mathbf{x}}) + \frac{1}{2}\left(\mathbf{c}^T(\mathbf{x})\mathbf{c}(\mathbf{x}) - \mathbf{c}^T(\bar{\mathbf{x}})\mathbf{c}(\bar{\mathbf{x}})\right)$$

$$+\frac{1}{2\gamma^2}\mathbf{w}(\mathbf{x}, \bar{\mathbf{x}})\mathbf{w}^T(\mathbf{x}, \bar{\mathbf{x}}) \leq 0 \tag{7.34}$$

$$\mathbf{w}(\mathbf{x}, \bar{\mathbf{x}}) = V_{\mathbf{x},\bar{\mathbf{x}}}(\mathbf{x}, \bar{\mathbf{x}})\bar{\mathbf{b}}(\mathbf{x}, \bar{\mathbf{x}}) \tag{7.35}$$

and there exists $\mathbf{l}(\bar{\mathbf{x}})$ solving

$$V_{\bar{\mathbf{x}}}(\mathbf{x}, \bar{\mathbf{x}})\mathbf{l}(\bar{\mathbf{x}}) = \mathbf{c}^T(\bar{\mathbf{x}}) \tag{7.36}$$

Then, it holds

$$\|\mathbf{z}_{\bar{\Sigma}}\|_2^2 \le \gamma^2 \|\mathbf{u}\|_2^2 + 2V(\mathbf{x}(0), \bar{\mathbf{x}}(0)). \tag{7.37}$$

Proof. Considering

$$\dot{V}(\mathbf{x}, \bar{\mathbf{x}}) = V_{\mathbf{x}, \bar{\mathbf{x}}}(\mathbf{x}, \bar{\mathbf{x}}) \left(\bar{\mathbf{a}}(\mathbf{x}, \bar{\mathbf{x}}) + \bar{\mathbf{b}}(\mathbf{x}, \bar{\mathbf{x}})\mathbf{u} \right) + V_{\bar{\mathbf{x}}}(\mathbf{x}, \bar{\mathbf{x}})\mathbf{l}(\bar{\mathbf{x}})\mathbf{z}_{\bar{\Sigma}}. \tag{7.38}$$

Note that

$$\frac{1}{2} \|\mathbf{z}_{\bar{\Sigma}}\|^2 = \frac{1}{2}\mathbf{y}^T\mathbf{y} + \frac{1}{2}\bar{\mathbf{y}}^T\bar{\mathbf{y}} - \bar{\mathbf{y}}^T\mathbf{y} \tag{7.39}$$

and moreover

$$\frac{1}{2} \left\| \gamma\mathbf{u} - \gamma^{-1}\mathbf{w}^T(\mathbf{x}, \bar{\mathbf{x}}) \right\|^2 =$$

$$\frac{\gamma^2}{2} \|\mathbf{u}\|^2 - \mathbf{w}(\mathbf{x}, \bar{\mathbf{x}})\mathbf{u} + \frac{1}{2\gamma^2}\mathbf{w}(\mathbf{x}, \bar{\mathbf{x}})\mathbf{w}^T(\mathbf{x}, \bar{\mathbf{x}}). \tag{7.40}$$

It turns out, by HJI (7.34) and (7.36),

$$\dot{V}(\mathbf{x}, \bar{\mathbf{x}}) = V_{\mathbf{x}, \bar{\mathbf{x}}}(\mathbf{x}, \bar{\mathbf{x}}) \left(\bar{\mathbf{a}}(\mathbf{x}, \bar{\mathbf{x}}) + \bar{\mathbf{b}}(\mathbf{x}, \bar{\mathbf{x}})\mathbf{u} \right) - \frac{1}{2} \left(\|\mathbf{z}_{\bar{\Sigma}}\|^2 - \|\mathbf{y}\|^2 + \|\bar{\mathbf{y}}\|^2 \right)$$

$$\le -\frac{1}{2} \left\| \gamma\mathbf{u} - \gamma^{-1}\mathbf{w}^T(\mathbf{x}, \bar{\mathbf{x}}) \right\|^2 + \frac{\gamma^2}{2} \|\mathbf{u}\|^2 - \frac{1}{2} \|\mathbf{z}_{\bar{\Sigma}}\|^2$$

$$\le \frac{\gamma^2}{2} \|\mathbf{u}\|^2 - \frac{1}{2} \|\mathbf{z}_{\bar{\Sigma}}\|^2 . \tag{7.41}$$

It follows from (7.41) that

$$V(\mathbf{x}(t), \hat{\mathbf{x}}(t)) - V(\mathbf{x}(0), \hat{\mathbf{x}}(0)) \le \int_0^t \frac{\gamma^2}{2}\|\mathbf{u}(\tau)\|^2 d\tau - \int_0^t \frac{\gamma^2}{2}\|\mathbf{z}_{\bar{\Sigma}}(\tau)\|^2 d\tau. \tag{7.42}$$

Thus, one has that

$$\|\mathbf{z}_{\bar{\Sigma}}\|_2^2 \le \gamma^2 \|\mathbf{u}\|_2^2 + 2V(\mathbf{x}(0), \hat{\mathbf{x}}(0)) \tag{7.43}$$

which completes the proof.   □

It is noted that in order to attain a more efficient FD performance, the evaluation window $[0, \tau]$ is adopted. Therefore, the following $\mathcal{L}_2$ fault detection scheme is proposed:

- Build the kernel representation $K_{\hat{\Sigma}}^{\bar{\mathbf{x}}_0}$ based on Theorem 7.3. As a result, in the fault-free case it holds that

$$\|(\mathbf{z}_{\bar{\Sigma}})_\tau\|_2^2 \leq \gamma^2 \|\mathbf{u}_\tau\|_2^2 + 2V(\mathbf{x}(0), \bar{\mathbf{x}}(0)) \qquad (7.44)$$

- Optimize fault diagnosis filter $\Sigma_{Q_f}^{\mathbf{x}_{s_0}}$ to get a better FD performance.

- Suppose the $\mathcal{L}_2$ gain of Σ_f is γ_σ, then the threshold can be set as

$$J_{\text{th}} = \gamma_\sigma^2(\gamma^2 \|\mathbf{u}_\tau\|_2^2 + 2V(\mathbf{x}(0), \bar{\mathbf{x}}(0))) \qquad (7.45)$$

- Design the detection logic

$$\begin{cases} J(\mathbf{r}) = \|\mathbf{r}_\tau\|_2^2 > J_{\text{th}} \implies \text{faulty} \\ J(\mathbf{r}) = \|\mathbf{r}_\tau\|_2^2 \leq J_{\text{th}} \implies \text{fault-free.} \end{cases} \qquad (7.46)$$

7.3.2 Controller Design

Next, the nominal controller is designed in the fault-free case to stabilize the plant as a prior. Concerning that the separation principle does not always hold in all nonlinear systems, there is still no generalized solution for the observer-based controller design. However, some researchers made the effort to circumvent this difficulty by imposing some constraints, see [138, 133]. In these schemes, the observer-based controller design can be achieved by

- determining a detectable SKR (7.12) to get the state estimation $\bar{\mathbf{x}}$.

- designing a control law $\mathbf{k}(\mathbf{x})$ to realize the internally stability of the feedback system

$$\begin{aligned} \dot{\mathbf{x}} &= \mathbf{a}(\mathbf{x}) + \mathbf{b}(\mathbf{x})\mathbf{k}(\mathbf{x}) \\ \mathbf{y} &= \mathbf{c}(\mathbf{x}) \end{aligned} \qquad (7.47)$$

- constructing the observer-based control law $\mathbf{k}(\bar{\mathbf{x}})$.

As the above fault diagnosis system also provides us the state estimation $\bar{\mathbf{x}}$, we only have to concentrate on determining the control law $\mathbf{k}(\mathbf{x})$.

Theorem 7.4. *Given the system (7.1). Suppose that there exists a solution $W(\mathbf{x}) \geq 0$ to the following Hamilton-Jacobi equation*

$$W_{\mathbf{x}}(\mathbf{x})\mathbf{a}(\mathbf{x}) - \frac{1}{2}W_{\mathbf{x}}(\mathbf{x})\mathbf{b}(\mathbf{x})\mathbf{b}^T(\mathbf{x})W_{\mathbf{x}}^T(\mathbf{x}) + \frac{1}{2}\mathbf{c}^T(\mathbf{x})\mathbf{c}(\mathbf{x}) = 0 \quad (7.48)$$

then $\mathbf{k}(\mathbf{x}) = -\mathbf{b}^T(\mathbf{x})W_{\mathbf{x}}^T(\mathbf{x})$ can internally stabilize the system (7.1).

Proof. The internal stability of (7.47) is equivalent to the stability of $K_{\{\Sigma,\Sigma_C\}}^{-1}$ for

$$\Sigma^{\mathbf{x}_0} : \begin{cases} \dot{\mathbf{x}} = \mathbf{a}(\mathbf{x}) + \mathbf{b}(\mathbf{x})\mathbf{u} \\ \mathbf{y} = \mathbf{c}(\mathbf{x}) \end{cases} \quad (7.49)$$

$$K^{\mathbf{x}_0} : \mathbf{u} = \mathbf{k}(\mathbf{x}). \quad (7.50)$$

A kernel representation of feedback system $\{\Sigma, \Sigma_C\}^{\mathbf{x}_0}$ is

$$K_{\{\Sigma,\Sigma_C\}}^{\mathbf{x}_0} : \begin{cases} \dot{\mathbf{x}} = \mathbf{f}(\mathbf{x}) + \mathbf{g}(\mathbf{x})\mathbf{u} + \mathbf{l}(\mathbf{x})(\mathbf{y} - \mathbf{h}(\mathbf{x})) \\ \mathbf{z}_\Sigma = \mathbf{y} - \mathbf{h}(\mathbf{x}) \\ \mathbf{z}_{\Sigma_C} = \mathbf{u} - \mathbf{k}(\mathbf{x}). \end{cases} \quad (7.51)$$

It is noted that the controller is first designed in the fault-free case. Then $\mathbf{z}_\Sigma = \mathbf{0}$ and $K_{\{\Sigma,\Sigma_C\}}^{-1}$ is

$$K_{\{\Sigma,\Sigma_C\}}^{-1} : \begin{cases} \dot{\mathbf{x}} = \mathbf{f}(\mathbf{x}) + \mathbf{g}(\mathbf{x})(\mathbf{z}_{\Sigma_C} + \mathbf{k}(\mathbf{x})) \\ \mathbf{y} = \mathbf{h}(\mathbf{x}) \\ \mathbf{u} = \mathbf{z}_{\Sigma_C} + \mathbf{k}(\mathbf{x}). \end{cases} \quad (7.52)$$

As can be referred to [134], if there exists a solution $W(\mathbf{x}) \geq 0$ to (7.48) and $\mathbf{k}(\mathbf{x}) = -\mathbf{b}^T(\mathbf{x})W_{\mathbf{x}}^T(\mathbf{x})$, then (7.52) is $\mathcal{L}_2$ stable. Furthermore $\mathbf{k}(\mathbf{x}) = -\mathbf{b}^T(\mathbf{x})W_{\mathbf{x}}^T(\mathbf{x})$ can internally stabilize the system. The proof is thus completed. □

Furthermore, by applying $\mathbf{u} = -\mathbf{b}^T(\bar{\mathbf{x}})W_{\bar{\mathbf{x}}}^T(\bar{\mathbf{x}})$, the design of the nominal controller is achieved. When an alarm is given by the fault diagnosis system, the residual-driven compensator will be employed to take care of the action caused by unknown fault variables, instead of redesigning the nominal controller. There are several schemes proposed for the compensator design: switching scheme, adaptive scheme and

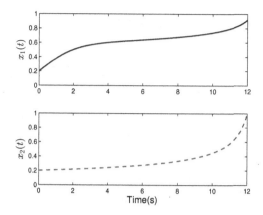

Figure 7.5: Time evolution of state without controller

on-line learning scheme, etc. For the switching scheme, $\Sigma_Q^{\mathbf{x}_q,0}$ or $\Sigma_{Q_d}^{\mathbf{x}_{q_d},0}$ represents a class of pre-designed fault-tolerant compensators while the fault diagnosis system serves as the switching parameters. One of the advantages is that the switching does not destabilize the control loop. As for the adaptive and on-line learning schemes, both of them can adjust the dynamic controller on-line to recover the system performance. Towards the design of the dynamic compensator, some work can be found, such as, in [129, 34, 165].

7.4 A Numerical Example

Consider the following nonlinear system

$$\Sigma : \begin{cases} \dot{x}_1(t) = -x_1^3(t) + x_2(t) \\ \dot{x}_2(t) = x_2^3(t) + u(t) + w(t) \\ y(t) = x_1(t) \end{cases}$$

where w is the fault signal. In this study, the initial condition is chosen as $\mathbf{x}(0) = \begin{bmatrix} 0.2 & 0.2 \end{bmatrix}^T$. The nominal controller design is first designed in the fault-free case. Note that the above nonlinear system is unstable, as depicted in Fig. 7.5. The nominal controller

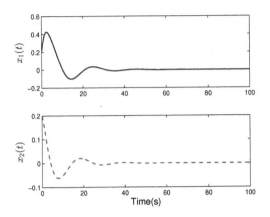

Figure 7.6: Time evolution of state with nominal controller Σ_C in fault-free case

$$\Sigma_C : \begin{cases} \dot{\bar{x}}_1(t) = -\bar{x}_1^3(t) + \bar{x}_2(t) - 13(y(t) - \bar{x}_1(t)) \\ \dot{\bar{x}}_2(t) = \bar{x}_2^3(t) - 7.5(y(t) - \bar{x}_1(t)) + u(t) \\ r(t) = y(t) - \bar{x}_1(t) \\ u(t) = -2\bar{x}_2^3(t) - 0.1\bar{x}_1(t) - 0.2\bar{x}_2(t) \end{cases}$$

is designed to stabilize the system as a priori with the initial condition $\bar{\mathbf{x}}(0) = \begin{bmatrix} 0 & 0 \end{bmatrix}^T$. As a result, the evolution of system state is shown in Fig. 7.6.

A constant fault $w = 0.05$ is simulated at $t = 100s$, which leads to the static deviation of the system state (see Fig. 7.7). A fault diagnosis unit is integrated to monitor the operation of the process with $J(r) = ||r(t)||$. The threshold is set based on Corollary 3.1. When an alarm is given (see Fig. 7.8), the dynamic controller

$$\Sigma_{C_Q} : \begin{cases} \dot{\bar{x}}_1(t) = -\bar{x}_1^3(t) + \bar{x}_2(t) - 13(y(t) - \bar{x}_1(t)) \\ \dot{\bar{x}}_2(t) = \bar{x}_2^3(t) - 7.5(y(t) - \bar{x}_1(t)) + u(t) \\ r(t) = y(t) - \bar{x}_1(t) \\ u(t) = -2\bar{x}_2^3(t) - 0.1\bar{x}_1(t) - 0.2\bar{x}_2(t) + \Sigma_Q(r(t)) \end{cases}$$

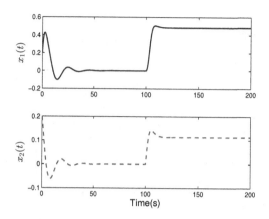

Figure 7.7: Time evolution of state with nominal controller Σ_C in faulty case

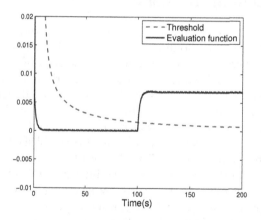

Figure 7.8: Fault detection system

is activated. The fault tolerance is attained with $\Sigma_Q(r(t)) = 10.4r(t)$ (as shown in Fig. 7.9). Thus, it can be concluded that the controller parametrization enables the separate design of system stability and fault tolerance.

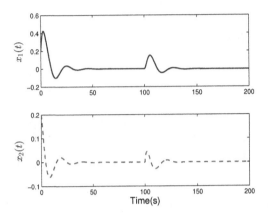

Figure 7.9: Time evolution of state with controller Σ_{C_Q} in faulty case

7.5 Concluding Remarks

The first part of this chapter has been devoted to investigate the FTC configurations for affine nonlinear systems. To this end, the observer and residual generator based controller parametrization, consists of an observer-based nominal controller and a residual-driven dynamic compensator, has been first introduced to attain the separate design of system performance and fault tolerance. Furthermore, an alternative controller framework composed by any stabilizing controller and a dynamic compensator has been developed, which has also established the relation between different types of controllers. It is worth noting that both of the controller frameworks emphasize the distinguished role of the residual access in the controller design. Thus, by involving these two types of residual generator based controller parametrization and an embedded fault diagnosis system, two FTC configurations have been proposed aiming at attaining the life-circle management for automatic control systems. All system parameters have been classified functionally with high and low priorities. In the second part of this work, one design scheme for the fault-tolerant configuration has been discussed by virtue of HJIs.

8 Application to Benchmark Processes

In this chapter, we will demonstrate the applications of the proposed FD and FTC schemes in the previous chapters to real processes. For this purpose, two laboratory benchmark processes are used. To be specific, the methods proposed in Chapters 4 and 6 are applied to the continuous stirred tank heater (CSTH) process, a typical nonlinear control system widely used in chemical process industries. The methods proposed in Chapters 7 and 5 are applied to the three-tank system.

8.1 Case Studies on CSTH Process

In this section, case studies on laboratory CSTH process are performed to demonstrate the effectiveness of the proposed $\mathcal{L}_2$ fuzzy observer-based FD systems for both continuous and discrete-time nonlinear processes.

8.1.1 Process Description

The laboratory setup CSTH plant considered here is a RT 682 agitator vessel experimental unit manufactured by G.U.N.T. Gerätebau GmbH Hamburg, as shown in Fig. 8.1. The schematic of the CSTH plant is sketched in Fig. 8.2. It facilitates experiments on an agitator vessel, which is heated by a heating jacket. During the experiment, water is continuously fed and conducted as the medium.

The main dynamics of the plant are the change of temperature as a function of the heat flows in and out of the tank, and the change of level as a function of in- and out-flowing water masses. Without considering the dynamics of the heat exchanger, the system dynamics can be represented by the tank volume V_T, the enthalpy in the tank H_T and the water

Figure 8.1: Laboratory setup of CSTH process

temperature in the heating jacket T_{hj} and modeled by

$$
\begin{bmatrix} \dot{V}_{\text{T}} \\ \dot{H}_{\text{T}} \\ \dot{T}_{\text{hj}} \end{bmatrix} = \begin{bmatrix} \dot{V}_{\text{in}} - \dot{V}_{\text{out}} \\ g\left(T_{\text{hj}} - \frac{H_{\text{T}}}{m_{\text{T}} \cdot c_p}\right) + \dot{m}_{\text{in}} c_p T_{\text{in}} - H_{\text{T}} \frac{\dot{V}_{\text{out}}}{V_{\text{T}}} \\ \frac{-g\left(T_{\text{hj}} - \frac{H_{\text{T}}}{m_{\text{T}} \cdot c_p}\right)}{m_{\text{hj}} \cdot c_p} + \frac{P_h}{m_{\text{hj}} \cdot c_p} \end{bmatrix}
$$

with $g\left(T_{\text{hj}} - \frac{H_{\text{T}}}{m_{\text{T}} \cdot c_p}\right)$ indicating some nonlinear function. After numerical tests, it has been shown that the nonlinear function $g\left(x_3 - \frac{x_2}{m_{\text{T}} \cdot c_p}\right)$ can be determined by a lookup table, as depicted in Fig. 8.3. The physical meaning of the process variables and parameters are listed in Table 8.1. Note that the tank itself is cylindrical with a base area $A_{\text{tank}} = \pi(100\text{mm})^2$ and a maximal height $h_{\text{T,max}} = 45\text{cm}$. Assembled inside the tank are an overflow pipe and the stirrer apparatus which takes up volume. It has been tested by experiment that the stirrer apparatus has a negligible effect on the volume, which leads to the linear relation between the water volume and level in the tank. Since the volume taken

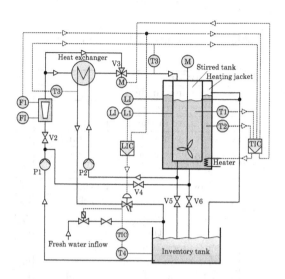

Figure 8.2: The schematic of CSTH plant [31]

up by the pipe can be calculated by $V_{\text{pipe}} = h_T(t) \cdot A_{\text{pipe}}$, the water volume can be described as a function of the measurable water level h_T as

$$V_T(t) = h_T(t) \cdot (A_{\text{tank}} - A_{\text{pipe}}) = h_T(t) \cdot A_{\text{eff}}.$$

The temperature of water in the tank T_T and water in the heating jacket T_{hj} can be measured directly. The relation between enthalpy H_T and temperature T_T is given as

$$T_T = \frac{H_T}{m_T c_p}.$$

Thus, h_T, T_{hj} and T_T are regarded as the measurable output variables.

The water inflow $\dot{V}_{\text{in}}$ is manipulated by a hand valve and thus considered to be constant during the operation while the output flow $\dot{V}_{\text{out}}$ can be regulated by human or controller. The heater can be either switched on or off and delivers a heating power of 1710W to the water when turned on. Therefore, $u_1 = \dot{V}_{\text{in}} - \dot{V}_{\text{out}}$ and $u_2 = P_h$ are considered to be the

Table 8.1: Technical data of CSTH process

Symbol	Description	Unit
V_T	water volume in the tank	L
H_T	enthalpy in the tank	J
T_{hj}	temperature in the heating jacket	°C
$\dot{V}_{in}, \dot{V}_{out}$	water flows in and out of the tank	l/s
$\dot{H}_{hj}$	enthalpy flow from the jacket to tank	J/s
$\dot{H}_{in}$	enthalpy flows from in-flowing water	J/s
$\dot{H}_{out}$	enthalpy flows from out-flowing water	J/s
m_{hj}	water mass in the heating jacket	kg
P_h	electrical heater power	J/s
h_T	water level in the tank	m
T_T	water temperature in the tank	°C
m_T	water mass in the tank	kg
$\dot{m}_{in}, \dot{m}_{out}$	mass flows in and out of the tank	kg/s
T_{in}	temperature of the in-flowing water	°C
T_{out}	temperature of the out-flowing water	°C
A_{eff}	the base area of the tank	m^2
c_p	heat capacity of water	J/kg°C

input variables. To sum up, by defining

$$
\mathbf{x} = \begin{bmatrix} x_1 \\ x_2 \\ x_3 \end{bmatrix} = \begin{bmatrix} V_T \\ H_T \\ T_{hj} \end{bmatrix}, \quad \mathbf{y} = \begin{bmatrix} y_1 \\ y_2 \\ y_3 \end{bmatrix} = \begin{bmatrix} h_T \\ T_T \\ T_{hj} \end{bmatrix}
$$

the following nonlinear model is obtained

$$
\dot{\mathbf{x}} = \begin{bmatrix} u_1 \\ g\left(x_3 - \frac{x_2}{m_T \cdot c_p}\right) + \dot{m}_{in} c_p T_{in} - \frac{\dot{V}_{out} x_2}{x_1} \\ \frac{-g\left(x_3 - \frac{x_2}{m_T \cdot c_p}\right)}{m_{hj} \cdot c_p} + \frac{u_2}{m_{hj} \cdot c_p} \end{bmatrix}, \mathbf{y} = \begin{bmatrix} \frac{x_1}{A_{eff}} \\ \frac{x_2}{c_p \cdot \rho \cdot x_1} \\ x_3 \end{bmatrix}
$$

where ρ represents the water-mass density which varies slightly with respect to the temperature.

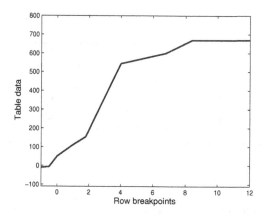

Figure 8.3: Lookup table for nonlinear function g

8.1.2 $\mathcal{L}_2$ Fuzzy Observer-based FD for Continuous Nonlinear Systems

We first show the effectiveness of the $\mathcal{L}_2$ fuzzy observer-based FD for continuous nonlinear systems proposed in Chapter 4.

In this case, two PI controllers

$$u_1(t) = \dot{V}_{\text{in}} - 0.017\left(v_1(t) - h_{\text{T}}(t)\right) - 10^{-4} \int_0^t \left(v_1(\tau) - h_{\text{T}}(\tau)\right) d\tau$$

$$u_2(t) = 0.01(v_2(t) - T_{\text{T}}(t)) + 10^{-8} \int_0^t (v_2(\tau) - T_{\text{T}}(\tau)) d\tau$$

are employed such that the process is running around the steady operation with $\mathbf{v}(t) = [v_1(t)\ v_2(t)]^T$ as the set points for the water level of the tank and the water temperature of the heating jacket, respectively. Note that the relation between u_1 and v_1 is also nonlinear, which leads the closed loop system to be a general type of nonlinear system as (4.1).

We first linearize the closed-loop system around the operation points listed in Table 8.2. It is noted that for convenience of presentation, we mainly concentrate on the change of the set point of water level in the tank at different operation points. Therefore, only the reference signal v_1 is considered as the premise variable. Moreover, the idea behind

Table 8.2: Operating points for CSTH process

Name	Value 1	2	3
Water level h_T / [cm]	10	20	30
Water temperature T_T / [°C]	26.6	26.6	26.6
Water temperature T_hj / [°C]	33.4	33.4	33.4

this premise variable setting role also lies in that the reference signal v_1 generally can not be influenced by external disturbances or fault signal, which promises a more efficient decision logic from the fuzzy set to the resulting inferred fuzzy system. Then, by taking into account the approximation errors between the original nonlinear model and the linearized models, the following T-S fuzzy model can be obtained:

Plant rule $\Re^i$: *IF* $v_1(t)$ *is* N_1^i

$$THEN \begin{cases} \dot{\mathbf{x}}(t) = \mathbf{A}_i\mathbf{x}(t) + \mathbf{B}_i\mathbf{v}(t) + \Delta_\mathbf{A}(\mathbf{x},\mathbf{v})\mathbf{x}(t) + \Delta_\mathbf{B}(\mathbf{x},\mathbf{v})\mathbf{v}(t) \\ \mathbf{y}(t) = \mathbf{C}_i\mathbf{x}(t) + \Delta_\mathbf{C}(\mathbf{x})\mathbf{x}(t), \quad i \in \{1,2,3\} \end{cases}$$

where

$$\mathbf{A}_1 = \begin{bmatrix} -0.0185 & 0 & 0 \\ 721.60 & -0.0107 & 43.2 \\ 0 & 0.000000136 & -0.0115 \end{bmatrix}$$

$$\mathbf{A}_2 = \begin{bmatrix} -0.0197 & 0 & 0 \\ 362.05 & -0.0053 & 43.2 \\ 0 & 0.00000006 & -0.0115 \end{bmatrix}$$

$$\mathbf{A}_3 = \begin{bmatrix} -0.0207 & 0 & 0 \\ 250.78 & -0.00375 & 43.2 \\ 0 & 0.000000047 & -0.0115 \end{bmatrix}$$

$$\mathbf{B}_1 = \begin{bmatrix} 0.00576 & 0 \\ 0 & 0 \\ 0 & 0.0101 \end{bmatrix}, \mathbf{B}_2 = \begin{bmatrix} 0.00611 & 0 \\ 0 & 0 \\ 0 & 0.0101 \end{bmatrix}$$

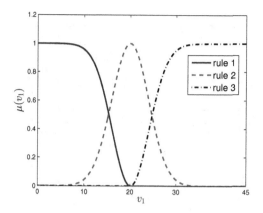

Figure 8.4: Membership functions for fuzzy modelling

$$
\mathbf{B}_3 = \begin{bmatrix} 0.00644 & 0 \\ 0 & 0 \\ 0 & 0.0101 \end{bmatrix}, \mathbf{C}_1 = \begin{bmatrix} 3.215 & 0 & 0 \\ 0 & 0.0000770 & 0 \\ 0 & 0 & 1 \end{bmatrix}
$$

$$
\mathbf{C}_2 = \begin{bmatrix} 3.215 & 0 & 0 \\ 0 & 0.0000385 & 0 \\ 0 & 0 & 1 \end{bmatrix}, \mathbf{C}_3 = \begin{bmatrix} 3.215 & 0 & 0 \\ 0 & 0.0000264 & 0 \\ 0 & 0 & 1 \end{bmatrix}.
$$

The membership functions for premise variable v_1 are depicted in Fig. 8.4. It is assumed that the model uncertainties $\Delta_\mathbf{A}(\mathbf{x}, \mathbf{v})$, $\Delta_\mathbf{B}(\mathbf{x}, \mathbf{v})$ and $\Delta_\mathbf{C}(\mathbf{x})$ have the upper bound as $\epsilon_1 = 0.001$, $\epsilon_2 = 0.0003$ and $\epsilon_3 = 0.0002$, respectively.

By applying Theorem 4.2, the minimum $\alpha = 0.00608$ is obtained with gain matrices

$$
\mathbf{L}_1 = \begin{bmatrix} 12.4404 & -7.0317 & 4.8500 \\ 54.6860 & 222.468 & -43.0303 \\ 4.4729 & -8.3943 & 31.3792 \end{bmatrix}
$$

$$
\mathbf{L}_2 = \begin{bmatrix} 11.2937 & -6.8069 & 2.1823 \\ 199.4145 & 122.8788 & 122.8797 \\ 1.2600 & -8.5437 & 31.4222 \end{bmatrix}
$$

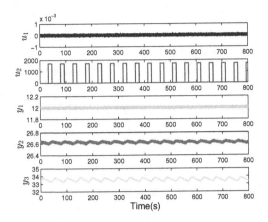

Figure 8.5: Input and output signals of the process in fault-free case

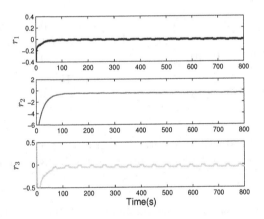

Figure 8.6: Residual signal $\mathbf{r}(t)$ in fault-free case

$$\mathbf{L}_3 = \begin{bmatrix} 10.6751 & -6.3301 & 1.0907 \\ 242.9536 & 128.6827 & 168.7900 \\ 0.2923 & -8.7632 & 31.3746 \end{bmatrix}.$$

To cope with real-world environment, all the output measured variables are set with the injection of noises. For demonstration purpose, the system

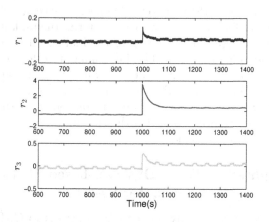

Figure 8.7: Residual signal $\mathbf{r}(t)$ in faulty case

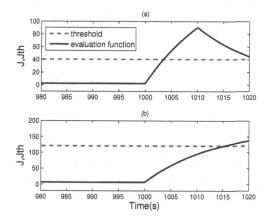

Figure 8.8: (a) Fault detection of a sensor fault for evaluation window $\tau = 10$s; (b) fault detection of a sensor fault for evaluation window $\tau = 400$s

is maintained in a steady operation as shown in Fig. 8.5. By carrying out the inferred fuzzy residual generator proposed in (4.10), it can be evidently seen from the simulation results depicted in Fig. 8.6 that the residual signal $\mathbf{r}(t)$ is $\mathcal{L}_2$ bounded in the fault-free case.

In order to illustrate the fault detection performance, a $-4°C$ offset of the measurement for the temperature of water in the tank is simulated at 1200s, which causes the deviation of residual signal, as depicted in Fig. 8.7. By adopting the residual evaluation and threshold computation method provided above with the evaluation window $\tau = 10$s and $\tau = 400$s respectively, it is evident that the fault can be detected in both cases, as shown in Fig. 8.8. By comparison, we can conclude that the detection delay can be significantly reduced with a proper evaluation window at hand in practical applications.

8.1.3 Weighted $\mathcal{L}_2$ Fuzzy Observer-based FD for Discrete-Time Nonlinear Systems

In what follows, a case study on the laboratory setup CSTH plant is utilized to illustrate the results of weighted $\mathcal{L}_2$ Fuzzy observer-based FD for discrete-time nonlinear systems proposed in Chapter 6.

To achieve a steady operation for the plant, two PI controllers

$$u_1(t) = \dot{V}_{in} - 0.0448\,(v_1(t) - h_T(t)) - 0.000264 \int_0^t (v_1(\tau) - h_T(\tau))\,d\tau$$

$$u_2(t) = 0.01(v_2(t) - T_T(t)) + 10^{-8} \int_0^t (v_2(\tau) - T_T(\tau))d\tau$$

are employed. For our purpose, we first linearize the closed-loop system around the operation points listed in Table 8.2 and then discretize it with sampling time $T = 5$s. Then, by taking into account of the approximation errors between the original nonlinear model and the linearized models, the following T-S fuzzy dynamic models can be obtained:

Plant rule $\Re^l$: *IF* $v_1(k)$ *is* N_1^l, *THEN*

$$\begin{cases} \mathbf{x}(k+1) = \mathbf{A}_l\mathbf{x}(k) + \mathbf{B}_l\mathbf{v}(k) + \Delta_\mathbf{A}(\mathbf{x}(k), \mathbf{v}(k))\mathbf{x}(k) \\ \qquad\qquad + \Delta_\mathbf{B}(\mathbf{x}(k), \mathbf{v}(k))\mathbf{v}(k) \\ \mathbf{y}(k) = \mathbf{C}_l\mathbf{x}(k) + \Delta_\mathbf{C}(\mathbf{x}(k))\mathbf{x}(k), \quad l \in \{1, 2, 3\} \end{cases}$$

where

$$\mathbf{A}_1 = \begin{bmatrix} 0.6572 & 0 & 0 \\ 1457.5 & 0.9737 & 207.1058 \\ 0.000254 & 0 & 0.9442 \end{bmatrix}, \mathbf{B}_1 = \begin{bmatrix} 0.1066 & 0 \\ 102.3029 & 5.3034 \\ 0 & 0.0491 \end{bmatrix}$$

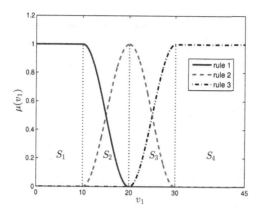

Figure 8.9: Membership functions for fuzzy modelling

$$\mathbf{A}_2 = \begin{bmatrix} 0.6572 & 0 & 0 \\ 2864.1 & 0.9482 & 204.3642 \\ 0.001 & 0 & 0.9442 \end{bmatrix}, \mathbf{B}_2 = \begin{bmatrix} 0.1066 & 0 \\ 202.0521 & 5.2568 \\ 0 & 0.0491 \end{bmatrix}$$

$$\mathbf{A}_3 = \begin{bmatrix} 0.6572 & 0 & 0 \\ 1013.8 & 0.9814 & 207.9348 \\ 0.000124 & 0 & 0.9441 \end{bmatrix}, \mathbf{B}_3 = \begin{bmatrix} 0.1066 & 0 \\ 71.0557 & 5.3175 \\ 0 & 0.0491 \end{bmatrix}$$

$$\mathbf{C}_1 = \begin{bmatrix} 3.21 & 0 & 0 \\ 0 & 0.0004 & 0 \\ 0 & 0 & 1 \end{bmatrix}, \mathbf{C}_2 = \begin{bmatrix} 3.21 & 0 & 0 \\ 0 & 0.00008 & 0 \\ 0 & 0 & 1 \end{bmatrix}$$

$$\mathbf{C}_3 = \begin{bmatrix} 3.21 & 0 & 0 \\ 0 & 0.00003 & 0 \\ 0 & 0 & 1 \end{bmatrix}.$$

The membership functions for premise variable v_1 are depicted in Fig. 8.9. Based on the proposed partition method, the following four regions are obtained: $\mathcal{S}_1 := \{v_1|0 < v_1 \leq 10\}, \mathcal{S}_2 := \{v_1|10 < v_1 \leq 20\}, \mathcal{S}_3 := \{v_1|20 < v_1 \leq 30\}, \mathcal{S}_4 := \{v_1|30 < v_1 \leq 45\}$. It is assumed that the model uncertainties $\Delta_{\mathbf{A}}(\mathbf{x},\mathbf{v})$, $\Delta_{\mathbf{B}}(\mathbf{x},\mathbf{v})$ and $\Delta_{\mathbf{C}}(\mathbf{x})$ have the upper bound as $\epsilon_1 = 0.0003$, $\epsilon_2 = 0.0001$ and $\epsilon_3 = 0.0001$, respectively.

It can be observed that the process is in a steady operation after adopting PI controllers (as shown in Fig. 8.10). To carry out the

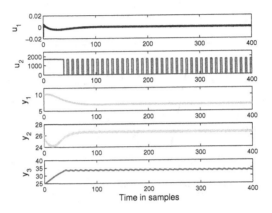

Figure 8.10: Input and output signals of the CSTH plant

weighted piecewise-fuzzy residual generator (6.35), the weighting factors are calculated according to Algorithm 3 first. As a result, one has that

$$\omega_1 = 9.49, \ \omega_2 = 9.16, \ \omega_3 = 9.16, \ \omega_4 = 10.37$$

and the gain matrices for the residual generator as

$$\mathbf{L}_{11} = \begin{bmatrix} 0.2278 & 0.0039 & 0.0145 \\ 853.7118 & 617.9733 & 216.4049 \\ 0.0376 & 0.0087 & 0.7848 \end{bmatrix}$$

$$\mathbf{L}_{21} = \begin{bmatrix} 0.2315 & 0.0157 & 0.0159 \\ 852.4627 & 550.3369 & 215.9024 \\ 0.0429 & 0.0121 & 0.7650 \end{bmatrix}$$

$$\mathbf{L}_{22} = \begin{bmatrix} 0.1941 & 0.0047 & 0.0051 \\ 444.8825 & 433.8545 & 218.8196 \\ 0.0093 & 0.0034 & 0.7712 \end{bmatrix}$$

$$\mathbf{L}_{32} = \begin{bmatrix} 0.2204 & 0.0075 & 0.0159 \\ 453.7491 & 326.2313 & 211.6489 \\ 0.0339 & 0.0055 & 0.7183 \end{bmatrix}$$

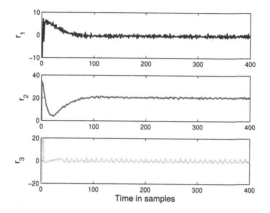

Figure 8.11: Residual signal in fault-free case

$$\mathbf{L}_{33} = \begin{bmatrix} 0.2213 & 0.0046 & 0.0233 \\ 323.2896 & 225.9315 & 201.7700 \\ 0.0394 & 0.0027 & 0.7046 \end{bmatrix}$$

$$\mathbf{L}_{43} = \begin{bmatrix} 0.2167 & 0.0037 & 0.0285 \\ 314.8217 & 202.0408 & 204.5818 \\ 0.0401 & 0.0034 & 0.6184 \end{bmatrix}.$$

As shown in Fig. 8.11, the residual signal is $\mathcal{L}_2$ bounded in fault-free case. To further demonstrate the effectiveness of the proposed FD scheme, a 4°C offset of the measurement for the temperature of water in the tank is simulated from 400th sample and removed from 500th sample. Thus, the corresponding residual signal in faulty case is given in Fig. 8.12. In addition, by adopting the on-line FD algorithm provided above with the evaluation window $\tau = 10$ samples, it can be evidently observed from Fig. 8.13 that the fault can be detected. Meanwhile, by adopting the evaluation window, it can be seen that the residual signals retain to the fault-free case and the alarm is released after removing the measurement fault. Moreover, suppose that the hand valve V1 is closed by about 50% to reduce the water inflow from 800th sample, the associated residual signals and a successful fault detection based on the FD algorithm are shown in Fig. 8.14 and Fig. 8.15, respectively.

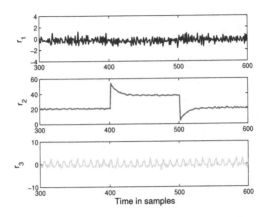

Figure 8.12: Residual signal in faulty case

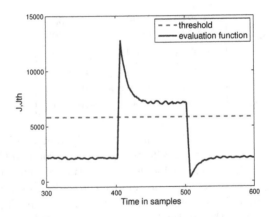

Figure 8.13: Detection of the measurement fault for the temperature of the tank

8.2 Case Studies on Three-Tank System

Next, case studies on the laboratory setup DTS200 of three-tank system are used to demonstrate the propose methods in Chapters 5 and 7.

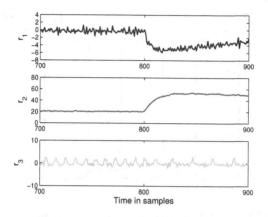

Figure 8.14: Residual signal in faulty case

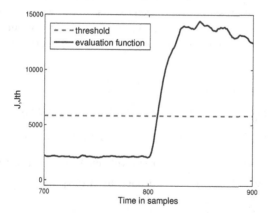

Figure 8.15: Detection of the actuator fault

8.2.1 Process Description

Three-tank system has typical characteristics of tanks, pipelines and pumps used in chemical processes. As sketched in Fig. 8.16, there are two pumps with incoming mass flow rate Q_1, Q_2, which pump water to

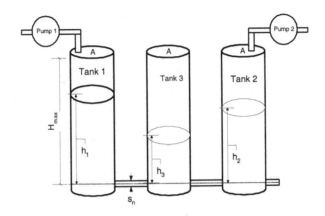

Figure 8.16: The schematic of laboratory three-tank system [31]

tank 1 and tank 2, respectively. Tank 3 is connected to tank 1 and tank 2 through pipes of cross sectional areas s_{13}, s_{23}. the water levels in tank 1 and tank 2 are measured through sensors. Applying the incoming and outgoing mass flows under consideration of Torricelli's law, the dynamics of three-tank system is modeled by

$$A\dot{h}_1 = Q_1 - a_1 s_{13}\mathrm{sgn}(h_1 - h_3)\sqrt{2g|h_1 - h_3|}$$
$$A\dot{h}_2 = Q_2 + a_3 s_{23}\mathrm{sgn}(h_3 - h_2)\sqrt{2g|h_3 - h_2|} - a_2 s_0\sqrt{2gh_2}$$
$$A\dot{h}_3 = a_1 s_{13}\mathrm{sgn}(h_1 - h_3)\sqrt{2g|h_1 - h_3|} - a_3 s_{23}\mathrm{sgn}(h_3 - h_2)\sqrt{2g|h_3 - h_2|}$$

where $h_i(t), i = 1, 2, 3$, are the water level (cm) in each tank and $s_{13} = s_{23} = s_0 = s_n$. $\mathbf{y} = \begin{bmatrix} h_1 & h_2 \end{bmatrix}^T$ are measured outputs. The parameters and variables are given in Table 8.3.

8.2.2 Real-Time Implementation of FTC Architecture with Embedded Runge-Kutta Iterations

In this study, we focus on real-time implementation of the FTC architecture for nonlinear systems in Fig. 7.3.

The core of the Youla parametrization of stabilizing controllers shown in Fig. 7.3 is the observer-based kernel representation [134], which is also called residual generator in FDD/FTC study [30]. As known

Table 8.3: Parameters of three-tank system [30]

Parameters	Symbol	Value	Unit
cross section area of tanks	$\mathcal{A}$	154	cm^2
cross section area of pipes	s_n	0.5	cm^2
max. height of tanks	$H_{\max}$	62	cm
max. flow rate of pump 1	$Q_{1_{\max}}$	100	cm^3/s
max. flow rate of pump 2	$Q_{2_{\max}}$	100	cm^3/s
coeff. of flow for pipe 1	a_1	0.475	
coeff. of flow for pipe 2	a_2	0.6	
coeff. of flow for pipe 3	a_3	0.475	

[134], in case of a stable nonlinear system, the process model itself can serve the construction of the kernel representation. In this case, the Youla parametrization of stabilizing controllers is reduced to an internal model control (IMC) configuration, as sketched in Fig. 8.17 (without the feedback of $\mathbf{r}$ and the fault estimate $\hat{\mathbf{f}}$). IMC has been widely used in engineering applications due to its significant role in reference tracking design and disturbance attenuation [58, 34]. In the IMC-based FTC architecture, the residual signal $\mathbf{r}(t) = \mathbf{z}(t) - \hat{\mathbf{z}}(t)$ is achieved by the real-time computation of the process model and plays a central role both for (feedback) control and FDD [34, 30]. $\Sigma_Q^{\mathbf{x}_q,0}$ represents a (stable) controller designed depending on the required control objectives. $\mathbf{w}$ is the reference signal. $\hat{\Sigma}^{\hat{\mathbf{x}}_0}$ denotes the process model which will be realized by iterative computation of the nonlinear model.

Multiplicative or parametric faults result in changes in system dynamics and can even cause system instability. In the context of IMC-based FTC, parametric faults lead to a mismatch between $\Sigma^{\mathbf{x}_0}$ and $\hat{\Sigma}^{\hat{\mathbf{x}}_0}$. An FTC scheme typically consists of a two-step procedure [164]: (a) fault detection and estimation (b) real-time adaptation of the process $\hat{\Sigma}^{\hat{\mathbf{x}}_0}$ using the fault estimates. Fault estimation can be approached both using parameter identification and observer-based techniques. It is the subject of many researchers [63, 57, 158, 54] and beyond the scope of dissertation. In the sequel, we focus on the second step, real-time computation of $\hat{\Sigma}^{\hat{\mathbf{x}}_0}$ based on the fault estimates to compensate the effects of the fault and ensure the system stability and performance. To this end, improved iterative computation methods will be applied to the real-time imple-

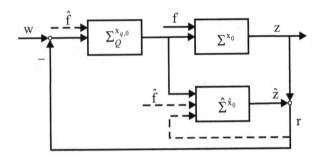

Figure 8.17: IMC with the embedded numerical iteration

mentation of $\hat{\Sigma}^{\hat{\mathbf{x}}_0}$, both in the faulty and fault-free cases. In order to meet demands for high real-time ability, we develop a modified numerical computation method [152], in which the available residual signal $\mathbf{r}(t)$ is fed back into the numerical iteration of $\hat{\Sigma}^{\hat{\mathbf{x}}_0}$, as shown in Fig. 8.17. This can significantly improve the convergence rate and computing accuracy of the iteration.

In this case study, by defining

$$\mathbf{x} = \begin{bmatrix} x_1 \\ x_2 \\ x_3 \end{bmatrix} = \begin{bmatrix} h_1 \\ h_2 \\ h_3 \end{bmatrix}, \mathbf{u} = \begin{bmatrix} u_1 \\ u_2 \end{bmatrix} = \begin{bmatrix} Q_1 \\ Q_2 \end{bmatrix}$$

nonlinear model is obtained

$$\dot{\mathbf{x}} = \phi(t, \mathbf{x}(t)) \tag{8.1}$$

$$\phi(t, \mathbf{x}(t)) = \begin{bmatrix} \dfrac{-a_1 s_{13}\mathrm{sgn}(x_1-x_3)\sqrt{2g|x_1-x_3|}}{\mathcal{A}} + \dfrac{u_1(t)}{\mathcal{A}} \\ \dfrac{a_3 s_{23}\mathrm{sgn}(x_3-x_2)\sqrt{2g|x_3-x_2|}-a_2 s_0\sqrt{2gh_2}}{\mathcal{A}} + \dfrac{u_2(t)}{\mathcal{A}} \\ \dfrac{a_1 s_{13}\mathrm{sgn}(x_1-x_3)\sqrt{2g|x_1-x_3|}-a_3 s_{23}\mathrm{sgn}(x_3-x_2)\sqrt{2g|x_3-x_2|}}{\mathcal{A}} \end{bmatrix}. \tag{8.2}$$

For our purpose, 4th order RK algorithm is first adopted to realize an observer-based residual generator with

$$\mathbf{A} = \begin{bmatrix} 0 & 0 & 0 & 0 \\ 0.5 & 0 & 0 & 0 \\ 0 & 0.5 & 0 & 0 \\ 0 & 0 & 1 & 0 \end{bmatrix}, \mathbf{b} = \begin{bmatrix} \frac{1}{6} \\ \frac{1}{3} \\ \frac{1}{3} \\ \frac{1}{6} \end{bmatrix}, \mathbf{c} = \begin{bmatrix} 0 \\ \frac{1}{2} \\ \frac{1}{2} \\ 1 \end{bmatrix}.$$

In [152], we study improving the iterative solutions of RK algorithm by embedding the sensor signals into the iterative computation. For iteratively solving the differential equation $\dot{\mathbf{x}} = \phi(t, x), \mathbf{x}(0) = \mathbf{x}_0$, we propose the following iterative computation algorithm

$$
\begin{aligned}
\mathbf{x}_{n+1} &= \mathbf{x}_n - \acute{\mathbf{B}}\varphi(t_n, \mathbf{y}_n) + \mathbf{L}(\mathbf{C}\mathbf{x}_n - z(n)) \\
\mathbf{y}_n &= \acute{\mathbf{C}}\mathbf{x}_n - \acute{\mathbf{D}}\varphi(t_n, \mathbf{y}_n)
\end{aligned}
\tag{8.3}
$$

where $\mathbf{L} \in \mathcal{R}^{k_x \times k_y}$ is the design parameter matrix to be determined. It is worth mentioning that (8.3) is of the observer structure, whose core is the feedback of signal $\mathbf{L}(\mathbf{C}\mathbf{x}_n - \mathbf{z}(n))$. In [152], we have shown that by a suitable selection of matrix $\mathbf{L}$, which is also called observer gain matrix, feeding back $\mathbf{L}(\mathbf{C}\mathbf{x}_n - \mathbf{z}(n))$ will significantly improve the convergence rate of the iteration. This scheme is called observer-based RK iterations.

In this study, the numerical iteration calculation starts with the initial condition $\mathbf{x}(0) = [10\ 5\ 12]^T$. The sector bounded condition is calculated as

$$
\begin{aligned}
\underline{\mathbf{K}} &= diag(-0.0633, -0.0649, -0.0649, -0.0665) \\
\overline{\mathbf{K}} &= diag(-0.002, -0.002, -0.002, -0.002).
\end{aligned}
$$

We first compare the convergence rate of the observer-like and the standard RK iterations applied to the three-tank benchmark. It is noted that the water level is at $h_1 = 30$cm, $h_2 = 20$cm by employing PI controller. In order to obtain a precise and real-time estimation of the system (8.1), $h = 0.48$ is adopted. Applying Theorem 5 in [152] to our system results in

$$
\mathbf{L} = \begin{bmatrix} 0.9831 & 0 \\ 0 & 0.9831 \\ 0 & 0 \end{bmatrix}
$$

with the minimum convergence rate $\varsigma = 2.1613$. As a result, the comparison of the RK iteration and the optimal observer-like RK iteration is demonstrated in Fig. 8.18. It costs around 3000 iterations for the standard RK algorithm to converge while observer-like RK only needs 300 iterations. Hence, the observer-like iteration can evidently improve the convergence rate of the RK estimation.

In the following, the IMC-based FTC with embedded iterative computation is studied by a simulation on the three-tank system, as shown

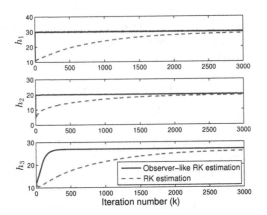

Figure 8.18: The comparison of RK estimation and observer-like RK estimation

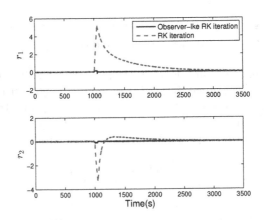

Figure 8.19: The comparison of estimation error between embedded RK iteration and observer-like RK iteration

in Fig. 8.19. Here $\Sigma_Q^{\mathbf{x}_q,0}$ is standard IMC controller which is realized by computing the inverse of the process model. $K_{\Sigma}^{\mathbf{\hat{x}}_0}$ is realized by RK iterations. By setting the reference signal as $w_1 = 45\text{cm}, w_2 = 15\text{cm}$ in

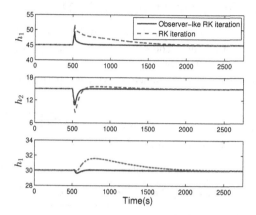

Figure 8.20: The comparison of FTC between embedded RK iteration and observer-like RK iteration

the IMC with embedded numerical iteration and observer-like numerical iteration, the output of the two systems will reach the operating point $h_1 = 45\text{cm}, h_2 = 15\text{cm}, h_3 = 30\text{cm}$ after some time.

It is noted that a fault estimation (FE) scheme is implemented which delivers the update of system parameters in real-time. For the FTC purpose, the pluggings between tank 1 and tank 3, tank 3 and tank 2 are simulated at $T = 1000$s, which exponentially changes the parameters of the pipes with $a_1 = a_3 = 0.475 - 0.175 \left(1 - e^{-0.03(t-T)}\right), t \geq T$. Once the faults are detected and estimated, the controller reconfiguration is realized by updating the parameters in RK iteration and controller $\Sigma_Q^{\mathbf{x}_q,0}$. Fig. 8.20 shows the comparison between the estimation error $\mathbf{r}(n) = \mathbf{C}\mathbf{x}_n - \mathbf{z}(n)$ of the observer-like and standard RK iterations after the occurrence of fault. It can be seen that the convergence rate of the residual signal is significantly improved with the embedded observer-like RK iteration. As a result, the fault-tolerant ability is enhanced, as demonstrated in Fig. 8.20. Overall, the real-time fault tolerance can be achieved with the improvement of the numerical iteration.

8.2.3 $\mathcal{L}_\infty/\mathcal{L}_2$ Fuzzy Observer-based FD

In what follows, a case study on the laboratory setup DTS200 of three-tank system, is used to show the effectiveness of $\mathcal{L}_\infty/\mathcal{L}_2$ fuzzy observer-based FD proposed in Chapter 5.

The three-tank system in presence of external disturbances is modeled by

$$
\begin{aligned}
\dot{x}_1 = {} & \mathcal{A}_{\text{inv}} u_1 - \mathcal{A}_{\text{inv}} a_1 s_{13} \text{sgn}(x_1 - x_3)\sqrt{2g|x_1 - x_3|} \\
& - \mathcal{A}_{\text{inv}} d_1 s_{13} \text{sgn}(x_1 - x_3)\sqrt{2g|x_1 - x_3|} \\
\dot{x}_2 = {} & \mathcal{A}_{\text{inv}} Q_2 + \mathcal{A}_{\text{inv}} a_3 s_{23}\text{sgn}(x_3 - x_2)\sqrt{2g|x_3 - x_2|} \\
& - \mathcal{A}_{\text{inv}} d_2 s_0 \sqrt{2gx_2} - \mathcal{A}_{\text{inv}} s_0 \sqrt{2gx_2} \\
\dot{x}_3 = {} & \mathcal{A}_{\text{inv}} a_1 s_{13}\text{sgn}(x_1 - x_3)\sqrt{2g|x_1 - x_3|} \\
& - \mathcal{A}_{\text{inv}} d_1 s_{13}\text{sgn}(x_1 - x_3)\sqrt{2g|x_1 - x_3|} \\
& - \mathcal{A}_{\text{inv}} a_3 s_{23}\text{sgn}(x_3 - x_2)\sqrt{2g|x_3 - x_2|}
\end{aligned}
$$

where $d_i(t), i = 1, 2$ represent the disturbances caused by pluggings between tank 1 and tank 3 and pluggings for the outflow of tank 2. The water level in tank 1 and tank 3 are measurable output variables.

For the purpose of demonstrating our results, the following T-S fuzzy models are used to approximate the dynamics of three-tank system first

Plant rule $\Re^i$: *IF* $x_1(t)$ *is* N_1^i, *THEN*

$$
\begin{cases}
\dot{x} = \mathbf{A}_i \mathbf{x} + \mathbf{B}_i \mathbf{u} + \mathbf{E}_i \mathbf{d} + \mathbf{a}_i + \Delta_\mathbf{A}(\mathbf{x})\mathbf{x} + \Delta_\mathbf{E}(\mathbf{x})\mathbf{d} \\
\mathbf{y} = \mathbf{C}_i \mathbf{x}, \quad i \in \{1, 2, 3\}
\end{cases}
$$

where

$$
\mathbf{A}_1 = \begin{bmatrix} -0.0221 & -0.0011 & 0.0191 \\ -0.0027 & -0.0359 & 0.0183 \\ 0.0205 & 0.0205 & -0.0409 \end{bmatrix}, \mathbf{a}_1 = \begin{bmatrix} 0.1 \\ 0.2 \\ 0 \end{bmatrix}
$$

$$
\mathbf{A}_2 = \begin{bmatrix} -0.0165 & -0.001 & 0.0130 \\ -0.0018 & -0.0290 & 0.0131 \\ 0.0145 & 0.0145 & -0.0289 \end{bmatrix}, \mathbf{a}_2 = \begin{bmatrix} 0 \\ 0 \\ 0 \end{bmatrix}
$$

$$
\mathbf{A}_3 = \begin{bmatrix} -0.0140 & -0.0009 & 0.0103 \\ -0.0012 & -0.0259 & 0.0110 \\ 0.0118 & 0.0118 & -0.0236 \end{bmatrix}, \mathbf{a}_3 = \begin{bmatrix} -0.1 \\ 0 \\ -0.2 \end{bmatrix}
$$

Table 8.4: Operating points for three-tank system

	Value		
Name	1	2	3
Water level h_1 / [cm]	15	10	12.5
Water level h_2 / [cm]	20	10	15
Water level h_3 / [cm]	25	10	12.5

$$
\mathbf{E}_1 = \begin{bmatrix} -0.3216 & 0 \\ 0 & -0.4548 \\ 0.3216 & 0 \end{bmatrix}, \mathbf{E}_2 = \begin{bmatrix} -0.4548 & 0 \\ 0 & -0.4548 \\ 0.4548 & 0 \end{bmatrix}
$$

$$
\mathbf{E}_3 = \begin{bmatrix} -0.5570 & 0 \\ 0 & -0.4548 \\ 0.5570 & 0 \end{bmatrix}, \mathbf{C}_1 = \mathbf{C}_2 = \mathbf{C}_3 = \begin{bmatrix} 1 & 0 & 0 \\ 0 & 1 & 0 \end{bmatrix}
$$

$$
\mathbf{B}_1 = \mathbf{B}_2 = \mathbf{B}_3 = \begin{bmatrix} 0.00649 & 0 \\ 0 & 0.00649 \\ 0 & 0 \end{bmatrix}.
$$

It is noted that the system parameters are obtained by linearizing the nonlinear system around the operation points listed in Tabel 8.4. For ease of presentation, we only focus on the change of the water level in tank 1. It is assumed that $||\Delta_\mathbf{A}(\mathbf{x})|| \leq 0.001$ and $||\Delta_\mathbf{E}(\mathbf{x})|| \leq 0.001$. The membership functions for premise variable $\theta(t) = x_3(t) - 15$ are given in Fig. 8.21 with $l_1 = 5, l_2 = 2.5, l_3 = 2.5, l_4 = 5$. Based on the partition method proposed in Section 5.2, the space for the premise variable can be divided into three regions, which are given by $S_1 := \{\theta| -l_1 < \theta \leq -l_2\}, S_2 := \{\theta| -l_2 < \theta \leq l_3\}, S_3 := \{\theta|l_3 < \theta \leq l_4\}$. Along the lines of [66], the constraint matrices can be constructed as follows:

$$
\vec{\mathbf{F}}_1 = \begin{bmatrix} -1 & 0 & 0 & -l_2 \\ 0 & 0 & 0 & 0 \\ 1 & 0 & 0 & 0 \\ 0 & 1 & 0 & 0 \\ 0 & 0 & 1 & 0 \end{bmatrix}, \mathbf{F}_2 = \begin{bmatrix} 0 & 0 & 0 \\ 0 & 0 & 0 \\ 1 & 0 & 0 \\ 0 & 1 & 0 \\ 0 & 0 & 1 \end{bmatrix}
$$

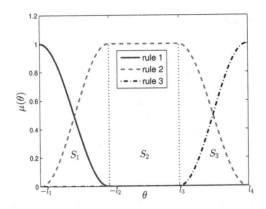

Figure 8.21: Membership functions for fuzzy modelling of three-tank system

$$\vec{\mathbf{F}}_3 = \begin{bmatrix} 0 & 0 & 0 & 0 \\ 1 & 0 & 0 & -l_3 \\ 1 & 0 & 0 & 0 \\ 0 & 1 & 0 & 0 \\ 0 & 0 & 1 & 0 \end{bmatrix}, \vec{\mathbf{H}}_1 = \begin{bmatrix} 1 & 0 & 0 & l_1 \\ -1 & 0 & 0 & -l_2 \\ 0 & 0 & 0 & 0 \end{bmatrix}$$

$$\mathbf{H}_2 = \begin{bmatrix} 0 & 0 & 0 \\ 0 & 0 & 0 \\ 0 & 0 & 0 \end{bmatrix}, \vec{\mathbf{H}}_3 = \begin{bmatrix} 1 & 0 & 0 & -l_3 \\ -1 & 0 & 0 & l_4 \\ 0 & 0 & 0 & 0 \end{bmatrix}.$$

By adopting the algorithm proposed in Theorem 5.2, we have $\alpha = 0.0035$ and

$$\mathbf{L}_{11} = \begin{bmatrix} 1.361 & 0 \\ 0 & 2.0432 \\ -0.2765 & 0 \end{bmatrix}, \mathbf{L}_{12} = \begin{bmatrix} 1.389 & 0 \\ 0 & 2.0162 \\ -0.2561 & 0 \end{bmatrix}$$

$$\mathbf{L}_{22} = \begin{bmatrix} 1.318 & 0 \\ 0 & 2.0710 \\ -0.3358 & 0 \end{bmatrix}, \mathbf{L}_{32} = \begin{bmatrix} 1.387 & 0 \\ 0 & 2.0152 \\ -0.2560 & 0 \end{bmatrix}$$

$$\mathbf{L}_{33} = \begin{bmatrix} 1.398 & 0 \\ 0 & 2.0352 \\ -0.2896 & 0 \end{bmatrix}.$$

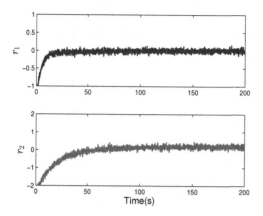

Figure 8.22: Residual signal in fault-free case

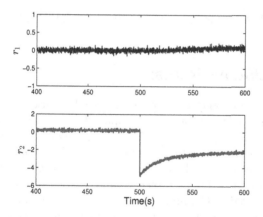

Figure 8.23: Residual signal in sensor fault case

For the demonstration purpose, disturbance $d_2 = 0.01 + 0.02\sin(t)$ is simulated. It is evident that the residual signal is $\mathcal{L}_2$-bounded in fault-free case, as shown in Fig. 8.22. In addition, an 4cm offset on the measurement of tank 2 is simulated at $t = 500$s. As a result, the corresponding residual signal is depicted in Fig. 8.23. It can be seen from

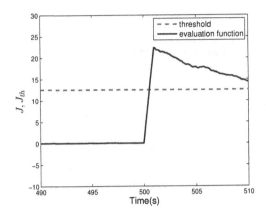

Figure 8.24: Detection of the sensor fault

Fig. 8.24 that the proposed scheme can realize a real-time detection of fault.

8.3 Concluding Remarks

In this chapter, two benchmark processes have been used to demonstrate the effectiveness of the proposed FD and FTC schemes. The fuzzy $\mathcal{L}_2$ observer-based FD approaches for both continuous and discrete-time systems proposed in Chapters 4 and 6 have been applied to CSTH process. In addition, the three-tank process has been used to show the effectiveness of the FTC architecture and $\mathcal{L}_\infty/\mathcal{L}_2$ observer-based FD approach proposed in Chapters 5 and 7, respectively. The results have shown that the proposed FD schemes can successfully detection faults and the proposed FTC architecture can be applied to attain fault tolerance. The results in this chapter provides the confidence in the applicability of the proposed design schemes in practice.

9 Conclusions and Future Work

In this thesis, the analysis and design issues of observer-based FD and FTC are addressed for nonlinear systems. In Chapter 1, the motivations and objectives of this work are presented. As nonlinearity is considered to exist in many practical systems, there appears an urgent and increasing demand for real-time FD and FTC for nonlinear systems. The parameterization forms of residual generators and controllers for LTI systems are summarized in Chapter 2, which play an essential role in formulating the linear FD and FTC configurations. Then, the state-of-the-art of the basic technology for nonlinear FD and FTC is also briefly reviewed.

Chapter 3 examines the essential analysis issues in nonlinear observer-based FD systems. Three different types ($\mathcal{L}_\infty, \mathcal{L}_2, \mathcal{L}_\infty/\mathcal{L}_2$ types) of FD systems are introduced, and, in this context, the associated existence and design conditions are studied. The proposed results are useful for the development of nonlinear FD systems using some well-established technologies. It is remarkable that the residual generator and evaluation as well as decision making are designed in an integrated manner. To further link the residual generation and evaluation as well as their optimization in the observer-based FD framework, a parameterization form of residual generators is studied. Based on the parameterization result, the FD systems with the threshold settings can be parameterized.

For the application of the results proposed in Chapter 3 to the FD system design, a mathematical and systematic tool is needed in order to to handle the nonlinear issues. To this end, the design schemes of $\mathcal{L}_2$ and $\mathcal{L}_\infty/\mathcal{L}_2$ types of observer-based FD systems are addressed with the aid of fuzzy dynamic modelling techniques and fuzzy/piecewise Lyapunov functions in Chapters 4 and 5, respectively. In particular, dynamic (adaptive) thresholds are proposed for the above two types of FD systems by taking into account of the influence of input variables on residual vectors, since unlike linear systems, the input variables of nonlinear systems will affect the residual signals. It is worth mentioning that the design approach proposed in Chapter 5 can be applied to the case that the premise variables of the FD system is non-synchronous with the premise variables of the fuzzy model for the plant.

Unlike linear systems with unified dynamics over the whole working range, the local behavior of nonlinear systems can be significantly different. From the practical viewpoint, the FD systems dealt with in Chapters 4 and 5 are designed in a manner of worst case handling of uncertainties. To further improve FD performance, a weighed fuzzy observer-based FD approach is proposed for nonlinear discrete-time systems in Chapter 6. The core of this approach is to make use of the knowledge provided by fuzzy models of each local region and then to weight the local residual signal by means of different weighting factors. This approach can lead to a significantly improvement of fault detectability.

With the FD system at hand, it is important to re-configure the controller to maintain or recover the system operations after an alarm is given. To this end, two FTC configurations are proposed for affine nonlinear systems in Chapter 7 with the integration of the fault diagnosis module. The design philosophy behind these two configurations begins with the parameterization of all stabilizing controllers in the observer-based form, and then proceeds with the formation of the expandable controller architecture with an integrated residual access. The proposed FTC architectures provide an integrated solution that has advantages to make the plant maintenance, repair and operations easier to handle.

Finally, the proposed approaches in Chapters 4-7 are tested on two benchmark processes as described in Chapter 8. The application results demonstrate the effectiveness of the developed methods.

Some of the future work are outlined as follows.

- The robustness issues of observer-based FD systems will be addressed for affine and more general types of nonlinear processes.

- The extension of the parameterization scheme of residual generators will be extended to nonlinear systems with disturbances, and based on it, the optimization of observer-based FD systems will be studied.

- The nonlinear observer-based FD systems will be further studied in the context of IOS and ROS [124].

- The observer-based realizations of controller parameterizations will be developed for more general types of nonlinear systems and, associated with it, the FTC configurations will be investigated.

Bibliography

[1] M. Abid, W. Chen, S. X. Ding, and A. Q. Khan, "Optimal residual evaluation for nonlinear systems using post-filter and threshold," *Int. J. Control*, vol. 84, no. 3, pp. 526–539, 2011.

[2] E. Alcorta-Garcia and P. M. Frank, "Deterministic nonlinear observer based approaches to fault diagnosis: A survey," *Control Engineering Practice*, vol. 5, no. 5, pp. 663–670, 1997.

[3] H. Alwi and C. Edwards, "Fault detection and fault-tolerant control of a civil aircraft using a sliding-mode-based scheme," *IEEE Trans. Control Syst. Technol.*, vol. 16, no. 3, pp. 499–510, 2008.

[4] H. Alwi, C. Edwards, and M. T. Hamayun, "Nonlinear integral sliding mode fault tolerant longitudinal aircraft control," in *IEEE International conference on Control Applications*, 2011, pp. 970–975.

[5] H. Alwi, C. Edwards, and C. P. Tan, *Fault Detection and Fault-Tolerant Control Using Sliding Modes*. Springer-Verlag, 2011.

[6] S. Aouaouda, M. Chadli, P. Shi, and H. R. Karimi, "Discrete-time H_-/H_∞ sensor fault detection observer design for nonlinear systems with parameter uncertainty," *Int. J. Robust Nonlinear Control*, vol. 25, no. 3, pp. 339–361, 2015.

[7] S. Armeni and A. C. C. Mosca, "Robust fault detection and isolation for LPV systems under a sensitivity constraint," *Int. J. Adapt. Control Signal Process.*, vol. 23, no. 1, pp. 55–72, 2009.

[8] Y.-H. Bae, S.-H. Lee, H.-C. Kim, B.-R. Lee, J. Jiang, and J. Lee, "A real-time intelligent multiple fault diagnostic system," *Int. J. Adv. Manuf Technol*, vol. 29, pp. 590–597, 2006.

[9] A. Banos, "Parameterization of nonlinear stabilizing controller-
     s: The observer-controller configuration," *IEEE Trans. Autom.
     Control*, vol. 43, no. 9, pp. 1268–1272, 1998.

[10] M. Basin, L. Li, M. Krueger, and S. X. Ding, "A finite-time-
     convergent fault-tolerant control for dynamical systems and its
     experimental verification for DTS200 three-tank system," *IET
     Control Theory & Applications*, 2015.

[11] M. Basseville and I. V. Nikiforov, *Detection of Abrupt Changes -
     Theory and Application.* Prentice-Hall, 1993.

[12] R. Beard, *Failure Accomondation in Linear Systems Through Self-
     Reorganization.* PhD dissertation, MIT, 1971.

[13] M. Benosman and K. Y. Lum, "Passive actuators' fault tolerant
     control for affine nonlinear systems," *IEEE Trans. Control Syst.
     Technol.*, vol. 18, no. 1, pp. 152–163, 2010.

[14] M. Blanke, M. Kinnaert, J. Lunze, and M. Staroswiecki, *Diagnosis
     and Fault-Tolerant Control, 2nd Edition.* Springer, 2006.

[15] J. Bokor and G. Balas, "Detection filter design for LPV systems -
     a geometric approach," *Automatica*, vol. 40, pp. 511–518, 2004.

[16] J. Bokor and Z. Szabo, "Fault detection and isolation in nonlinear
     systems," *Annual Reviews in Control*, vol. 33, no. 2, pp. 113–123,
     2009.

[17] F. Caccavale, F. Pierri, and L. Villani, "Adaptive observer for fault
     aidgnosis in nonlinear discrete-time systems," *Jounal of dynam-
     ic systems, measurement, and control*, vol. 130, pp. 021 005–1 –
     021 005–9, 2008.

[18] S. G. Cao, N. W. Rees, and G. Feng, "Analysis and design for
     a class of complex control systems Part I: fuzzy modelling and
     identification," *Automatica*, vol. 33, no. 6, pp. 1017–1028, 1997.

[19] P. Castaldi, N. Mimmo, and S. Simani, "Differential geometry
     based active fault tolerant control for aircraft," *Control Engineering
     Practice*, 2014.

[20] M. Chadli, A. Abdo, and S. X. Ding, "H_-/H_∞ fault detection filter design for discrete-time Takagi-Sugeno fuzzy systems," *Automatica*, vol. 49, no. 7, pp. 1996–2005, 2013.

[21] M. Chadli and T. M. Guerra, "LMI solution for robust static output feedback control of discrete TakagiSugeno fuzzy models," *IEEE Trans. Fuzzy Syst.*, vol. 20, no. 6, pp. 1160–1165, 2012.

[22] M. Chadli and H. R. Karimi, "Robust observer design for unknown inputs Takagi-Sugeno models," *IEEE Trans. Fuzzy Syst.*, vol. 21, no. 1, pp. 158–164, 2013.

[23] M. Chadli, D. Maquin, and J. Ragot, "Stability analysis and design for continuous-time Takagi-Sugeno control systems," *Int. J. Fuzzy Syst.*, vol. 7, no. 3, pp. 101–109, 2005.

[24] G. Chen and R. J. P. de Figueiredo, "Construction of the left coprime factional representation for a class of nonlinear control systems," *Syst. Control Lett.*, vol. 12, no. 4, pp. 353–361, 1990.

[25] J. Chen and R. J. Patton, *Robust Model-Based Fault Diagnosis for Dynamic Systems*. Kluwer Academic Publishers, 1999.

[26] S.-S. Chen, Y.-C. Chang, S.-F. Su, S.-L. Chung, and T.-T. Lee, "Robust static output-feedback stabilization for nonlinear discrete-time systems with time-delay via fuzzy control approach," *IEEE Trans. Fuzzy Syst.*, vol. 13, no. 2, pp. 263–272, 2005.

[27] W. Chen and M. Saif, "A sliding mode observer-based strategy for fault detection, isolation, and estimation in a class of Lipschitz nonlinear systems," *Int. J. Syst.. Science*, vol. 38, pp. 943–955, 2007.

[28] Y. Diao and K. M. Passino, "Stable fault-tolerant adaptive fuzzy/neural control for a turbine engine," *IEEE Trans. Control Syst. Technol.*, vol. 9, no. 3, pp. 494–509, 2001.

[29] ——, "Intelligent fault-tolerant control using adaptive and learning methods," *Control Engineering Practice*, vol. 10, pp. 801–817, 2002.

[30] S. X. Ding, *Model-Based Fault Diagnosis Techniques - Design Schemes, Algorithms and Tools, 2nd Edition*. London: Springer-Verlag, 2013.

[31] ——, *Data-Driven Design of Fault Diagnosis and Fault-Tolerant Control Systems.* Springer-Verlag, London, 2014.

[32] S. X. Ding, P. M. Frank, E. L. Ding, and T. Jeinsch, "A unified approach to the optimization of fault detection systems," *Int. J. Adapt. Control and Signal Processing*, vol. 14, pp. 725–745, 2000.

[33] S. X. Ding, B. Shen, Z. Wang, and M. Zhong, "A fault detection schme for linear discrete-time systems with an integrated online performance evaluation," *Int. J. Control*, vol. 87, no. 12, pp. 2511–2521, 2014.

[34] S. X. Ding, G. Yang, P. Zhang, E. Ding, T. Jeinsch, N. Weinhold, and M. Schulalbers, "Feedback control structures, embedded residual signals and feedback control schemes with an integrated residual access," *IEEE Trans. Control Syst. Technol.*, vol. 18, pp. 352–367, 2010.

[35] X. Ding and P. M. Frank, "Fault detection via factorization approach," *Syst. Control Lett.*, vol. 14, no. 5, pp. 431–436, 1990.

[36] ——, "An adaptive observer-based fault detection scheme for nonlinear systems," in *Proceedings of the 12th IFAC World Congress*, Sydney, 1993, pp. 307–312.

[37] H. Dong, Z. D. Wang, J. Lam, and H. Gao, "Fuzzy-model-based robust fault detection with stochastic mixed time delays and successive packet dropouts," *IEEE Trans. Syst., Man, Cybern., Part B. Cybern.*, vol. 42, no. 2, pp. 365–376, 2012.

[38] A. Edelmayer, J. Bokor, Z. Szabo, and F. Szigeti, "Input reconstruction by means of system inversion: A geometric approach to fault detection and isolation in nonlinear systems," *Int. J. Appl. Math. Comput. Sci.*, vol. 14, no. 2, pp. 189–199, 2004.

[39] C. Edwards, K. Spurgeon, and R. J. Patton, "Sliding mode observers for fault detection and isolation," *Automatica*, vol. 36, no. 4, pp. 541–553, 2000.

[40] G. Feng, *Analysis and Synthesis of Fuzzy Control Systems - A Model Based Approach.* Boca Raton, FL: CRC, 2010.

[41] G. Feng, C. Chen, D. Sun, and Y. Zhu, "H_∞ controller synthesis of fuzzy dynamic systems based on piecewise Lyapunov functions and bilinear matrix inequalities," *IEEE Trans. Fuzzy Syst.*, vol. 13, no. 1, pp. 94–103, 2005.

[42] T. Floquet, J. P. Barbot, W. Perruquetti, and M. Djemai, "On the robust fault detection via a sliding mode disturbance observer," *Int. J. Control*, vol. 77, no. 7, pp. 622–629, 2004.

[43] P. M. Frank and X. Ding, "Survey of robust residual generation and evaluation methods in observer-based fault detection systems," *J. Process Control*, vol. 7(6), pp. 403–424, 1997.

[44] K. Fujimoto, "Synthesis and analysis of nonlinear control systems based on transformations and factorizations," Ph.D. dissertation, Kyoto University, Japan, 2000.

[45] K. Fujimoto and T. Sugie, "State-space characterization of youla parametrization for nonlinear systems based on input-to-state stability," *Proc. of the 37th IEEE Conference on Decision and Control*, pp. 2479–2484, 1998.

[46] ——, "Characterization of all nonlinear stabilizing controllers via observer-based kernel representations," *Automatica*, vol. 36, no. 8, pp. 1123–1135, 2000.

[47] Q. Gao, G. Feng, Y. Wang, and J. Qiu, "Universal fuzzy controllers based on generalized T-S fuzzy models," *Fuzzy Sets and Systems*, vol. 201, pp. 55–70, 2012.

[48] Q. Gao, X. Zeng, G. Feng, Y. Wang, and J. Qiu, "T-S-fuzzy-model-based approximation and controller design for general nonlinear systems," *IEEE Trans. Syst., Man, Cybern., Part B. Cybern.*, vol. 42, no. 4, pp. 1143–1154, 2012.

[49] Z. Gao, X. Shi, and S. X. Ding, "Fuzzy state/disturbance observer design for T-S fuzzy systems with application to sensor fault estimation," *IEEE Trans. Syst., Man, Cybern., Part B. Cybern.*, vol. 38, no. 3, pp. 875–880, 2008.

[50] L. E. Ghaoui, F. Oustry, and M. AitRami, "A cone complementarity linearization algorithm for static output-feedback and related problems," *IEEE Trans. Autom. Control*, vol. 42, no. 8, pp. 1171–1176, 1997.

[51] F. Gustafsson, *Adaptive Filtering and Change Detection*. John Wiley and Sons, LTD, 2000.

[52] H. Hammouri, M. Kinnaert, and E. E. Yaagoubi, "Observer-based approach to fault detection and isolation for nonlinear systems," *IEEE Trans. Autom. Control*, vol. 44, no. 10, pp. 1879–1884, 1999.

[53] H. Hao, "Key performance monitoring and diagnosis in industrial automation processes," Ph.D. dissertation, University of Duisburg-Essen, 2014.

[54] X. He, Z. Wang, X. Wang, and D. H. Zhou, "Networked strong tracking filtering with multiple packet dropouts: algorithms and applications," *IEEE Trans. Ind. Electron.*, vol. 61, no. 3, pp. 1454–1463, 2014.

[55] X. He, Z. Wang, and D. Zhou, "Robust H_∞ filtering for time-delay systems with probabilistic sensor faults," *IEEE Signal Process. Lett.*, vol. 16, no. 5, pp. 442–445, 2009.

[56] J. Imura and T. Yoshikawa, "Parametrization of all stabilizing controllers of nonlinear systems," *Syst. Control Lett.*, vol. 29, no. 4, pp. 207–213, 1997.

[57] R. Isermann, *Fault Diagnosis Systems*. Springer-Verlag, 2006.

[58] A. Isidori, L. Marconi, and A. Serrani, *Robust Autonomous Guidance: An Internal Model Approach*. Springer, 2003.

[59] B. Jiang, Z. Gao, P. Shi, and Y. Xu, "Adaptive fault-tolerant tracking control of near-space vehicel using Takagi-Sugeno fuzzy models," *IEEE Trans. Fuzzy Syst.*, vol. 18, no. 5, pp. 1000–1007, 2010.

[60] B. Jiang, Z. Mao, and P. Shi, "H_∞-filter design for a class of networked control systems via T-S fuzzy model approach," *IEEE Trans. Fuzzy Syst.*, vol. 18, no. 1, pp. 201–208, 2010.

[61] B. Jiang, M. Staroswiecki, and V. Cocquempot, "Fault diagnosis based on adaptive observer for a class of non-linear systems with unknown parameters," *Int. J. Control*, vol. 77, pp. 415–426, 2004.

[62] ——, "Fault estimation in nonlinear uncertain systems using robust sliding-mode observers," *IEE Proceedings, Part D: Control Theory and Applications*, vol. 151, no. 1, pp. 29–37, 2004.

[63] ——, "Fault estimation in nonlinear uncertain systems using robust/sliding observers," *IEE Proc. Control Theory Appl.*, vol. 151, no. 1, pp. 29–37, 2004.

[64] ——, "Fault accommodation for nonlinear dynamic systems," *IEEE Trans. Autom. Control*, vol. 51, no. 9, pp. 1578–1583, 2006.

[65] M. Johansson, *Piecewise Linear Control Systems-A Computational Approach*. Springe-Verlag, 2002.

[66] M. Johansson, A. Rantzer, and K. Arzen, "Piecewise quadratic stability of fuzzy systems," *IEEE Trans. Fuzzy Syst.*, vol. 7, no. 6, pp. 713–722, 1999.

[67] H. Jones, *Failure Detection in Linear Systems*. PhD dissertation, MIT, 1973.

[68] A. C. Juan, "An algorithm to find minimal cuts of coherent fault-trees with event-classes." *IEEE Trans. Reliability*, vol. 48, no. 1, pp. 31–41, 1999.

[69] P. Kabore and H. Wang, "Design of fault diagnosis filters and fault-tolerant control for a class of nonlinear systems," *IEEE Trans. Autom. Control*, vol. 46, no. 11, pp. 1805–1810, 2001.

[70] E. Kamal, A. Aitouche, R. Ghorbani, and M. Bayart, "Robust fuzzy fault-tolerant control of wind energy conversion systems subject to sensor faults," *IEEE Trans. Sustain. Energy*, vol. 3, no. 2, pp. 231–241, 2012.

[71] I. Karafyllis and C. Kravaris, "Global exponential observers for two classes of nonlinear systems," *Syst. Control Lett.*, vol. 61, no. 7, pp. 797–806, 2012.

[72] A. Q. Khan, M. Abid, and S. X. Ding, "Fault detection filter design for discrete-time nonlinear systems - a mixed H_-/H_∞ optimization," *Syst. Control Lett.*, vol. 67, pp. 46–54, 2014.

[73] A. Q. Khan and S. X. Ding, "Threshold computation for fault detection in a class of discrete-time nonlinear systems," *Int. J. Adapt. Control Signal Process.*, vol. 25, no. 5, pp. 407–429, 2011.

[74] E. Kim and H. Lee, "New approaches to relaxed quadratic stability condition of fuzzy control systems," *IEEE Trans. Fuzzy Syst.*, vol. 8, no. 5, pp. 523–534, 2000.

[75] P. Korba, R. Babuška, H. B. Verbruggen, and P. M. Frank, "Fuzzy gain scheduling: Controller and observer design based on Lyapunov method and convex optimization," *IEEE Trans. Fuzzy Syst.*, vol. 11, no. 3, pp. 285–298, 2003.

[76] K. Kumamaru, J. Hu, K. Inoue, and T. Soederstroem, "Robust fault detection using index of kullback discrimination information," in *Proceedings of IFAC World Congress*, San Francisco, USA, 1996, pp. 205–210.

[77] Z. Lendek, T. Guerra, R. Babuska, and B. D. Schutter, *Stability Analysis and Nonlinear Observer Design using Takagi-Sugeno Fuzzy Models*. Netherlands: Springer-Verlag, 2010.

[78] Z. Lendek, J. Lauber, T. M. Guerra, R. Babuska, and B. D. Schutter, "Adaptive observers for TS fuzzy systems with unknown polynomial inputs," *Fuzzy Sets and Systems*, vol. 161, no. 15, pp. 2043–2065, 2010.

[79] G. Li, S. J. Qin, and D. Zhou, "Geometric properties of partial least squares for process monitoring," *Automatica*, vol. 46, no. 1, pp. 204–210, 2010.

[80] L. Li, S. X. Ding, J. Qiu, and Y. Yang, "Nonlinear observer-based fault detection approach and its T-S fuzzy-based implementation with application to industrial processes," *Preprint submitted to IEEE Trans. Syst., Man, Cybern., Part A. Syst.*, 2015.

[81] ——, "Real-time fault detection approach for nonlinear systems and its asynchronous T-S fuzzy observer-based implementation," *Accepted to IEEE Trans. Cybern.*, 2015.

[82] L. Li, S. X. Ding, J. Qiu, Y. Yang, and Y. Zhang, "Weighted fuzzy observer-based fault detection approach for discrete-time nonlinear systems via piecewise-fuzzy lyapunov functions," *Accepted to IEEE Trans. Fuzzy Syst.*, 2015.

[83] L. Li, S. X. Ding, Y. Yang, and Y. Zhang, "Robust fuzzy observer-based fault detection for nonlinear systems with disturbances," *Neurocomputing*, 2015.

[84] L. Li, Y. Yang, S. X. Ding, Y. Zhang, and S. Zhai, "On fault-tolerant control configurations for a class of nonlinear systems," *J. Franklin Institute*, vol. 352, no. 4, pp. 1397–1416, 2015.

[85] Y. Liang, D. Liaw, and T. Lee, "Reliable control of nonlinear systems," *IEEE Trans. Autom. Control*, vol. 45, no. 4, pp. 706–710, 2000.

[86] M. Liu and P. Shi, "Sensor fault estimation and tolerant control for ito stochastic systems with a descriptor sliding mode approach," *Automatica*, vol. 49, no. 5, pp. 1242–1250, 2013.

[87] M. Liu, P. Shi, L. Zhang, and X. Zhao, "Fault-tolerant control for nonlinear Markovian jump systems via proportional and derivative sliding mode observer technique," *IEEE Trans. Circuits Syst.*, vol. 58, no. 11, pp. 2755–2764, 2011.

[88] X. Liu, H. Zhang, J. Liu, and J. Yang, "Fault detection and diagnosis of permanent-magnet DC motor based on parameter estimation and neural network," *IEEE Trans. Ind. Electron.*, vol. 47, no. 5, pp. 1021–1030, 2000.

[89] X. Liu and Q. Zhang, "New approches to H_∞ controller design based on fuzzy observers for T-S fuzzy systems via LMI," *Automatica*, vol. 39, no. 9, pp. 1571–1582, 2003.

[90] Y. Liu, J. Wan, and G. Yang, "Reliable control of uncertain nonlinear systems," *Automatica*, vol. 34, no. 7, pp. 875–879, 1998.

[91] W.-M. Lu, "A state-space approach to parameterization of stabilizing controllers for nonlinear systems," *IEEE Trans. Autom. Control*, vol. 40, no. 9, pp. 1576–1588, 1995.

[92] D. G. Luenberger, "Observing the state of a linear system," *IEEE Trans. Mil. Electron*, vol. 8, no. 2, pp. 74–80, 1964.

[93] M. Mahmoud, J. Jiang, and Y. Zhang, *Active Fault Tolerant Control Systems*. Springer, 2003.

[94] R. Mangoubi, *Robust Estimation and Failure Detection*. Springer, 1998.

[95] Z. Mao, B. Jiang, and S. X. Ding, "A fault-tolerant control framework for a class of non-linear networked control systems," *Int. J. Syst. Sci.*, vol. 40, pp. 449–460, 2009.

[96] Z. Mao, B. Jiang, and P. Shi, "Observer-based fault-tolerant control for a class of networked control systems with transfer delays," *J. Franklin Institute*, vol. 348, pp. 763–776, 2011.

[97] M. A. Massoumnia, "A geometric approach to the synthesis of failure detection filters," *IEEE Trans. Autom. Control*, vol. AC-31, no. 9, pp. 839–846, 1986.

[98] N. Meskin and K. Khorasani, "Robust fault detection and isolation of time-delay systems using a geometric approach," *Automatica*, vol. 45, no. 6, pp. 1567–1573, 2009.

[99] C. Nan, F. Khan, and M. T. Iqbal, "Real-time fault diagnosis using knowledge-based expert system," *Process Safety and Environmental Protection*, vol. 86, no. 1, pp. 55–71, 2008.

[100] S.-K. Nguang, "H_∞ filtering for fuzzy dynamical systems with D stability constraints," *IEEE Trans. Circuits Syst. I, Fundam. Theory Appl*, vol. 50, no. 11, pp. 1503–1508, 2003.

[101] S.-K. Nguang and P. Shi, "H_∞ fuzzy output feedback control design for nonlinear systems: An LMI approach," *IEEE Trans. Fuzzy Syst.*, vol. 11, no. 3, pp. 331–340, 2003.

[102] S. K. Nguang, P. Shi, and S. X. Ding, "Fault detection for uncertain fuzzy systems: an LMI approach," *IEEE Trans. Fuzzy Syst.*, vol. 15, no. 6, pp. 1251–1262, 2007.

[103] J. O'Reilly, *Observers for Linear Systems*. London: Academic press, 1983.

[104] A. D. B. Paice and J. B. Moore, "On the Youla-Kucera parametrization for nonlinear systems," *Syst. Contr. Lett.*, vol. 14, no. 2, pp. 121–129, 1990.

[105] A. D. B. Paice and A. van der Schaft, "The class of stabilizing nonlinear plant controller pairs," *IEEE Trans. Autom. Control*, vol. 41, no. 5, pp. 634–645, 1996.

[106] A. Paice and A. V. der Schaft, "Stable kernel representations as nonlinear left coprime factorizations," *Proc. of the 33rd IEEE Conference on Decision and Control*, pp. 2786–2791, 1994.

[107] D.-J. Pan, Z.-Z. Han, and Z.-J. Zhang, "Bounded-input-boundd-output stabilization of nonlinear systems using state detectors," *Syst. Control Lett.*, vol. 21, no. 3, pp. 189–198, 1993.

[108] R. J. Patton, "Fault-tolerant control: The 1997 situation," *Proc. of the IFAC Symp. SAFEPROCESS 97*, pp. 1033–1055, 1997.

[109] C. D. Persis and A. Isidori, "A geometric approach to nonlinear fault detection and isolation," *IEEE Trans. Autom. Control*, vol. 46, no. 6, pp. 853–865, 2001.

[110] A. M. Pertew, H. J. Marquez, and Q. Zhao, "LMI-based sensor fault diagnosis for nonlinear Lipschitz systems," *Automatica*, vol. 43, no. 8, pp. 1464–1469, 2007.

[111] J. Qiu, G. Feng, and H. Gao, "Fuzzy-model-based piecewise H_{infty} static-output-feedback controller design for networked nonlinear systems," *IEEE Trans. Fuzzy Syst.*, vol. 18, no. 5, pp. 919–934, 2010.

[112] ——, "Asynchronous output-feedback control of networked nonlinear systems with multiple packet dropout: T-S fuzzy affine model-based approach," *IEEE Trans. Fuzzy Syst.*, vol. 19, no. 6, pp. 1014–1030, 2011.

[113] ——, "Observer-based piecewise affine output feedback controller synthesis of continuous-time T-S fuzzy affine dynamic systems using quantized measurements," *IEEE Trans. Fuzzy Syst.*, vol. 20, no. 6, pp. 1046–1062, 2012.

[114] ——, "Static-output-feedback H_∞ control of continuous-time T-S fuzzy affine systems via piecewise Lyapunov functions," *IEEE Trans. Fuzzy Syst.*, vol. 21, no. 2, pp. 245–261, 2013.

[115] Z. Qu, C. M. Ihlefeld, Y. Jin, and A. Saengdeejing, "A. robust fault-tolerant self-recovering control of nonlinear uncertain systems," *Automatica*, vol. 39, no. 10, pp. 1763–1771, 2003.

[116] T. J. Ross, *Fuzzy Logic with Engineering Applications.* Wiley, 1995.

[117] E. L. Russell, L. Chiang, and R. D. Braatz, *Data-driven Techniques for Fault Detection and Diagnosis in Chemical Processes.* London: Springer-Verlag, 2000.

[118] U. Shaked, "Improved LMI representations for the analysis and the design of continuous-time systems with polytopic type uncertainty," *IEEE Trans. Autom. Control*, vol. 46, no. 4, pp. 652–656, 2001.

[119] J. S. Shamma and M. Athans, "Guaranteed properties of gain scheduled control for linear parameter-varying plants," *Automatica*, vol. 27, no. 3, pp. 559–564, 1991.

[120] Q. Shen, B. Jiang, and V. Cocquempot, "Fault-tolerant control for T-S fuzzy systems with application to near-space hypersonic vehicel with actuator faults," *IEEE Trans. Fuzzy Syst.*, vol. 20, no. 4, pp. 652–665, 2012.

[121] P. Shi, Y. Yin, and F. Liu, "Gain-scheduled worst case control on nonlinear stochastic systems subject to actuator saturation and unknown information," *J. Optimization Theory and Applications*, vol. 156, no. 3, pp. 844–858, 2013.

[122] D. Shin and Y. Kim, "Reconfigurable flight control system design using adaptive neural networks," *IEEE Trans. Control Syst. Technol.*, vol. 12, no. 1, pp. 87–100, 2004.

[123] E. Sontag, "Smooth stabilization implies coprime factorization," *IEEE Trans. AC*, vol. 34, pp. 435–443, 1989.

[124] E. Sontag and Y. Wang, "Lyapunov characterizations of input to output stability," *SIAM. J. Control. Optim*, vol. 39, no. 1, pp. 226–249, 2001.

[125] D. J. Stilwell and W. J. Rugh, "Interpolation of observer state feedback controllers for gain scheduling," *IEEE Trans. Autom. Control*, vol. 44, no. 6, pp. 1225–1229, 1999.

[126] T. Takagi and M. Sugeno, "Fuzzy identification of systems and its applications to modeling and control," *IEEE Trans. Syst., Man, Cybern.-Part B: Cybern.*, vol. SMC-15, no. 1, pp. 116–132, 1985.

[127] C. Tan and C. Edwards, "Sliding mode observers for detection and reconstruction of sensor faults," *Automatica*, vol. 38, pp. 1815–1821, 2002.

[128] K. Tanaka, T. Ikeda, and H. O. Wang, "Fuzzy regulators and fuzzy observers: relaxed stability conditions and LMI-based design," *IEEE Trans. Fuzzy Syst.*, vol. 6, no. 2, pp. 250–265, 1998.

[129] T. T. Tay, I. M. Y. Mareels, and J. B. Moore, *High Performance Control.* Birkhuser, 1998.

[130] T. T. Tay and J. B. Moore, "Left coprime factorizattion and a class of stabilizing controllers for nonlinear systems," *Int. J. Control*, vol. 49, no. 4, pp. 1235–1248, 1989.

[131] M. C. M. Teixeira and S. H. Żak, "Stabilizing controller design for uncertain nonlinear systems using fuzzy models," *IEEE Trans. Fuzzy Syst.*, vol. 7, no. 2, pp. 133–142, 1999.

[132] S. C. Tong, B. Huo, and Y. . Li, "Observer-based adaptive decentralized fuzzy fault-tolerant control of nonlinear large-scale systems with actuator failures," *IEEE Trans. Fuzzy Syst.*, vol. 22, no. 1, pp. 1–15, 2014.

[133] J. Tsinias, "A generalization of Vidyasagar's theorem on stabilizability using state detection," *Syst. Control Lett.*, vol. 17, no. 1, pp. 37–42, 1991.

[134] A. van der Schaft, $\mathcal{L}_2$-*Gain and Passivity Techniques in Nonlinear Control.* Springer, 2000.

[135] V. Venkatasubramanian, R. Rengaswamy, K. Yin, and S. Kavuri, "A review of process fault detection and diagnosis part I: Quantitative model-based methods," *Computers and Chemical Engineering*, vol. 27, pp. 293–311, 2003.

[136] M. S. Verma, "Coprime fractional representations and stability of nonlinear feedback systems," *Int. J. Control*, vol. 48, no. 3, pp. 897–918, 1988.

[137] V. Verma, G. Gordon, R. Simmons, and S. Thrun, "Real-time fault diagnosis," *IEEE Robotics & Automation Magazine*, pp. 56–66, 2004.

[138] M. Vidyasagar, "On the stabilization of nonlinear systems using state detection," *IEEE Trans. Autom. Control*, vol. 25, no. 3, pp. 504–509, 1980.

[139] ——, *Control System Synthesis: A Factorization Approach.* MIT Press, Cambridge, 1985.

[140] H. Wang, Z. J. Huang, and S. Daley, "On the use of adaptive updating rules for actuator and sensor fault diagnosis," *Automatica*, vol. 33, pp. 217–225, 1997.

[141] H. Wang and G.-H. Yang, "Integrated fault detection and control for LPV systems," *Int. J. of Robust and Nonlinear Control*, vol. 19, pp. 341–363, 2009.

[142] G. Wei, G. Feng, and Z. Wang, "Robust H_∞ control for discrete-time fuzzy systems with infinite-distributed delays," *IEEE Trans. Fuzzy Syst.*, vol. 17, no. 1, pp. 224–232, 2009.

[143] M. Witczak, *Fault Diagnosis and Fault-Tolerant Control Strategies for Non-Linear Systems.* Springer-Verlag, 2014.

[144] L. Wu and W. C. Ho, "Fuzzy filter design for Itô stochastic systems with application to sensor fault detection," *IEEE Trans. Fuzzy Syst.*, vol. 17, no. 1, pp. 233–242, 2009.

[145] A. Xu and Q. Zhang, "Nonlinear system fault diagnosis based on adaptive estimation," *Automatica*, vol. 40, no. 7, pp. 1181–1193, 2004.

[146] A. V. Yablokov, V. B. Nesterenko, and A. V. Nesterenko, *Chernobyl: Consequences of the Catastrophe for People and the Environment.* Wiley-Blackwell, 2009.

[147] X. G. Yan and C. Edwards, "Nonlinear robust fault reconstruction and estimation using a sliding mode observer," *Automatica*, vol. 43, no. 9, pp. 1605–1614, 2007.

[148] X.-G. Yan and C. Edwards, "Robust decentralized actuator fault detection and estimation for large-scale systems using a sliding mode observer," *Int. J. Control*, vol. 81, no. 4, pp. 591–606, 2008.

[149] X. G. Yan and C. Edwards, "Robust sliding mode observer-based actuator fault detection and isolation for a class of nonlinear systems," *Int. J. Syst. Sci.*, vol. 39, no. 4, pp. 349–359, 2008.

[150] G. Yang, J. Lam, and J. Wang, "Reliable H_∞ control for affine nonlinear systems," *IEEE Trans. Autom. Control*, vol. 43, no. 8, pp. 1112–1117, 1998.

[151] Y. Yang, S. X. Ding, and L. Li, "On observer-based fault detection for nonlinear systems," *Syst. Control Lett.*, vol. 82, pp. 18–25, 2015.

[152] Y. Yang, L. Li, and S. X. Ding, "A control-theoretic study on Runge-Kutta methods with application to real-time fault-tolerant control of nonlinear systems," *IEEE Trans. Ind. Electron.*, vol. 62, no. 6, pp. 3914–3922, 2015.

[153] Y. Yang, Y. Zhang, S. X. Ding, and L. Li, "Design and implementation of lifecycle management for industrial control applications," *IEEE Trans. Control Syst. Technol.*, vol. 23, no. 4, pp. 1399–1410, 2015.

[154] Y. Yin, P. Shi, and C. Liu, "Gain-scheduled robust fault detection on time-delay stochastic nonlinear systems," *IEEE Trans. Ind. Electron.*, vol. 58, no. 10, pp. 4908–4916, 2011.

[155] X. Zeng, J. A. Keane, and D. Wang, "Fuzzy systems approach to approximation and stabilization of conventional affine nonlinear systems," in *Proceedings of the 2006 IEEE International Conference on Fuzzy Systems*, Vancouver, BC, Canada, 2006, pp. 277–284.

[156] X. Zeng and M. G. Singh, "Approximation theory of fuzzy systems-SISO case," *IEEE Trans. Fuzzy Syst.*, vol. 2, no. 2, pp. 162–176, 1994.

[157] H. Zhang, C. Li, and X. Liao, "Stability analysis and H_∞ controller design of fuzzy large-scale systems based on piecewise Lyapunov functions," *IEEE Trans. Syst., Man, Cybern., Part B. Cybern.*, vol. 36, no. 3, pp. 685–698, 2006.

[158] K. Zhang, B. Jiang, and P. Shi, "Observer-based integrated robust fault estimation and accommodation design for discrete-time systems," *Int. J. Control*, vol. 83, no. 6, pp. 1167–1181, 2010.

[159] ——, "Fault estimation observer design for discrete-time Takagi-Sugeno fuzzy systems based on piecewise Lyapunov functions," *IEEE Trans. Fuzzy Syst.*, vol. 20, no. 1, pp. 192–200, 2012.

[160] P. Zhang and S. X. Ding, "An integrated trade-off design of observer based fault detection systems," *Automatica*, vol. 44, pp. 1886–1894, 2008.

[161] Q. Zhang, "Adaptive observer for multiple-input-multiple-output (MIMO) linear time-varying systems," *IEEE Trans. Autom. Control*, vol. 47, pp. 525–529, 2002.

[162] X. Zhang, T. Parisini, and M. M. Polycarpou, "Adaptive fault tolerant control of nonlinear uncertain systems: An information based diagnostic approach," *IEEE Trans. Autom. Control*, vol. 49, no. 8, pp. 1259–1274, 2004.

[163] X. Zhang, M. M. Polycarpou, and T. Parisini, "Fault diagnosis of a class of nonlinear uncertain systems with Lipschitz nonlinearities using adaptive estimation," *Automatica*, vol. 46, no. 2, pp. 290–299, 2010.

[164] Y. Zhang and J. Jiang, "Bibliographical review on reconfigurable fault-tolerant control systems," *Annual Review in Control*, vol. 32, no. 2, pp. 229–252, 2008.

[165] Y. Zhang, Y. Yang, S. X. Ding, and L. Li, "Data-driven design and optimization of feedback control systems for industrial applications," *IEEE Trans. Ind. Electron.*, vol. 61, no. 11, pp. 6409–6417, 2014.

[166] Y. Zhao, J. Lam, and H. Gao, "Fault detection for fuzzy systems with intermittent measurements," *IEEE Trans. Fuzzy Syst.*, vol. 17, no. 2, pp. 398–410, 2009.

[167] Y. Zheng, H. Fang, and H. O. Wang, "Takagi-Sugeno fuzzy-model-based fault detection for networked control systems with Markov delays," *IEEE Trans. Syst., Man, Cybern., Part B. Cybern.*, vol. 36, no. 4, pp. 924–929, 2006.

[168] M. Zhong, S. Ding, J. Lam, and H. Wang, "An LMI approach to design robust fault detection filter for uncertain LTI systems," *Automatica*, vol. 39, pp. 543–550, 2003.

[169] K. Zhou, J. Doyle, and K. Glover, *Robust and Optimal Control*. Prentice Hall, 1996.

[170] K. Zhou and J. C. Doyle, *Essentials of Robust Control*. Upper Saddle River, NJ: Prience-Hall, 1998.

[171] K. Zhou and Z. Ren, "A new controller architecture for high performance, robust, and fault-tolerant control," *IEEE Trans. Autom. Control*, vol. 46, pp. 1613–1618, 2001.

Printed in the United States
By Bookmasters